Occupational Noise and Workplace Acoustics

Occupational Safety, Health, and Ergonomics: Theory and Practice

Series Editor: Danuta Koradecka
(Central Institute for Labour Protection – National Research Institute)

This series will contain monographs, references, and professional books on a compendium of knowledge in the interdisciplinary area of environmental engineering, which covers ergonomics and safety and the protection of human health in the working environment. Its aim consists in an interdisciplinary, comprehensive and modern approach to hazards, not only those already present in the working environment, but also those related to the expected changes in new technologies and work organizations. The series aims to acquaint both researchers and practitioners with the latest research in occupational safety and ergonomics. The public, who want to improve their own or their family's safety, and the protection of heath will find it helpful, too. Thus, individual books in this series present both a scientific approach to problems and suggest practical solutions; they are offered in response to the actual needs of companies, enterprises, and institutions.

Individual and Occupational Determinants: Work Ability in People with Health Problems
Joanna Bugajska, Teresa Makowiec-Dąbrowska, Tomasz Kostka

Healthy Worker and Healthy Organization: A Resource-Based Approach
Dorota Żołnierczyk-Zreda

Emotional Labour in Work with Patients and Clients: Effects and Recommendations for Recovery
Dorota Żołnierczyk-Zreda

New Opportunities and Challenges in Occupational Safety and Health Management
Daniel Podgórski

Emerging Chemical Risks in the Work Environment
Małgorzata Pośniak

Visual and Non-Visual Effects of Light: Working Environment and Well-Being
Agnieszka Wolska, Dariusz Sawicki, Małgorzata Tafil-Klawe

Occupational Noise and Workplace Acoustics: Advances in Measurement and Assessment Techniques
Edited by Dariusz Pleban

Virtual Reality and Virtual Environments: A Tool for Improving Occupational Safety and Health
Andrzej Grabowski

Head, Eye, and Face Personal Protective Equipment: New Trends, Practice and Applications
Katarzyna Majchrzycka

Nanoaerosols, Air Filtering and Respiratory Protection: Science and Practice
Katarzyna Majchrzycka

Microbial Corrosion of Buildings: A Guide to Detection, Health Hazards, and Mitigation
Rafał L. Górny

Respiratory Protection Against Hazardous Biological Agents
Katarzyna Majchrzycka, Justyna Szulc, Małgorzata Okrasa

For more information about this series, please visit: https://www.crcpress.com/Occupational-Safety-Health-and-Ergonomics-Theory-and-Practice/book-series/CRCOSHETP

Occupational Noise and Workplace Acoustics

Advances in Measurement and Assessment Techniques

Edited by
Dariusz Pleban

CRC Press
Taylor & Francis Group
Boca Raton London New York

CRC Press is an imprint of the
Taylor & Francis Group, an **informa** business

First edition published 2021
by CRC Press
6000 Broken Sound Parkway NW, Suite 300, Boca Raton, FL 33487-2742

and by CRC Press
2 Park Square, Milton Park, Abingdon, Oxon, OX14 4RN

ISBN: 9780367499259 (hbk)
ISBN: 9781003048121 (ebk)
ISBN: 9780367499280 (pbk)

Typeset in Times
by Lumina Datamatics Limited

Contents

Preface

Sound is defined as a wave disturbance propagating in an elastic medium and inducing acoustic vibrations of frequencies which can trigger an aural sensation in humans with normal hearing. Airborne sounds (such as speech, singing, music, or noise) are integral components of human's everyday life. Any undesired, unpleasant, annoying, onerous, or harmful sounds having an influence on the organ of hearing and other senses as well as other human body parts are commonly described as *noise*. In view of the omnipresence of noise (which surrounds modern man both in everyday life and in the work environment), it is necessary not only to aspire to minimize its effect on people but also to provide them with suitable acoustic comfort. Techniques and methods for noise emission control and minimization of its effect on humans have been developed and mastered over decades. These well-known and commonly practiced noise control measures were already discussed many times in monographs and textbooks. However, thanks to the continuous progress in the area of technology, it is possible to discuss and introduce into practice several new techniques and methods in the scope of noise measurement, assessment, and control.

This monograph offers the readers the opportunity to make themselves acquainted with the subject matter of selected up-to-date techniques and methods in the area of noise control and improvement of acoustic comfort, including techniques and methods developed in the Central Institute for Labour Protection—National Research Institute (CIOP-PIB). It comprises the latest knowledge concerning innovative advanced techniques of examination and evaluation of parameters characterizing the sound field and noise in the work environment, acoustic properties of rooms, and noise control measures. The range of issues discussed in this monograph encompasses acoustic field visualization techniques, methods of surround sound field reproduction in laboratory conditions, wireless sensor networks and the Internet of Things, methods of solving optimization problems with the use of genetic algorithms, room acoustic quality assessment methods, and measuring techniques specific for airborne ultrasounds (ultrasonic noise). The monograph strives to identify potential fields of effective application of these techniques and methods to improve the acoustic climate in the work environment.

The present monograph is divided into ten chapters. Chapter 1 is a review of basic concepts and quantities characterizing the phenomenon of sound which the reader will come across further in this book.

Sound field *visualization techniques* (the scanning method, the beamforming method, and the acoustic holography method) make up the subject matter of Chapter 2. Thanks to these techniques, one can "see" the sound field, and "snapshots" of sound can be used in research on noise emitted by machines and devices, offering thus the simplest possible tool for checking which element, constituting a component of a mechanism (very complex in some cases), contributes the most to the observed overall noise emission.

Chapter 3 is dedicated to issues pertaining to reproduction of the sound field in three dimensions. The proposed methodology of the creation of virtual *sound environments* not only enables the transfer of the physical location of the research work from "in situ" conditions to a laboratory but also significantly improves the quality of studies on the effect of noise on humans.

Aspirations to build human-friendly intelligent cities or the needs specific for Industry 4.0 result in increasing interest in *wireless sensor networks* (WSNs) and the Internet of Things as versatile tools that can be used to monitor, measure, and assess parameters characterizing work and everyday life environments. Applications of WSNs in the area of noise control are discussed in Chapter 4.

Up-to-date optimization techniques based on application of *genetic algorithms* are presented in Chapter 5. The potential of the genetic algorithm method in solving complex multi-dimensional optimization problems is presented on the example of a computer application which determines optimum spatial arrangement of noise sources and workstations in a workroom in order to minimize worker's exposure to noise.

The next two chapters of this monograph pertain to issues concerning assessment of the *acoustic quality of rooms* and measures aimed at bringing them into conformity with the requirements of applicable standards. The multi-index method presented in Chapter 6 offers the possibility of assessing the acoustic quality of classrooms with the use of a single-number global acoustic quality index. The method offers a comprehensive approach to the issue of acoustic assessment of classrooms, guaranteeing at the same time the uniqueness and comparability of the assessment results. Specific acoustic requirements are applicable for work in office rooms of the "open-plan" type. In Chapter 7, issues concerning parameters, research methods, and assessment criteria of global acoustic properties of *open-plan office rooms* are discussed and a methodology is recommended to be applied when adaptation of a room to applicable requirements (called *acoustic treatment*) is carried out.

The issues connected with measurements of airborne ultrasounds (i.e., the ultrasonic noise) in the work environment are the subject of the next two chapters. In the range of frequencies above 20 kHz, any unambiguous and in-depth information about factors affecting results of sound pressure level measurements is hardly available. None of the currently applicable international standards concerns methodology of taking *measurements of ultrasonic noise* at workstations in industrial conditions. Hence, a measuring method specifically adapted to the ultrasonic range of frequencies is presented in Chapter 8, together with an in-depth discussion of requirements concerning the measuring apparatus, recommended ways of conduct in the course of taking measurements, application of corrections to measurement results, and compilation of the budget of uncertainties. Chapter 9 is devoted to real-world practice of ultrasonic noise measurements and presents the methodology for evaluation of *sound-insulating properties of enclosures* for machines and devices emitting ultrasonic noise in the frequency range of 10–40 kHz.

One innovative and promising noise control technique consists in the use of *phononic crystals* to construct sound barriers. Chapter 10, devoted to this sound reduction method, includes description of computing and measuring techniques used in the research on phononic crystals, discussion of strengths and weaknesses of application of standardized measuring methods to studies on structures of that type, and presentation of results of the chapter author's own studies on the possibility of using sound field visualization techniques in assessment of effectiveness of sound barriers constructed in the form of phononic crystals.

All the chapters are illustrated with numerous examples of technical solutions, research setups, and results of measurement realized with the presented techniques, methods, and methodologies. Many of the techniques and solutions presented in this monograph, in view of their interdisciplinary nature, may be found interesting and useful by representatives of various fields of research and scientific disciplines other than acoustics.

Dariusz Pleban

MATLAB® is a registered trademark of The MathWorks, Inc. For product information, please contact:

The MathWorks, Inc.
3 Apple Hill Drive
Natick, MA 01760-2098 USA
Tel: 508 647 7000
Fax: 508-647-7001
E-mail: info@mathworks.com
Web: www.mathworks.com

Acknowledgment

We would like to thank all those without whom we would not have been able to complete and publish this monograph.

We offer our special thanks to Professor Wiktor M. Zawieska for his creative inspiration, patient encouragement, and invaluable scientific support in realization of research projects results of which are presented throughout the pages of this monograph.

We would like also to recognize the invaluable cooperation and assistance offered to us by our colleagues from the CIOP-PIB's Department of Vibroacoustic Hazards in carrying out experiments essential in the context of topics raised in this monograph.

Our grateful thanks are also extended to Dr. Jan K. Snakowski for his help in translating and editing the manuscript.

The Authors

Editor

Dariusz Pleban, BEng, PhD, DSc, is a professor at the Central Institute for Labour Protection—National Research Institute in Warsaw, Poland (CIOP-PIB), head of the Department of Vibroacoustic Hazards, and secretary of the CIOP-PIB Scientific Council. In his research work, he deals primarily with issues related to research and assessment methods of machinery and equipment noise emissions. His publishing achievements comprise more than 140 written works, including peer-reviewed articles in scientific journals and conference materials and compilations of a monographic nature. He is a member of the Committee on Acoustics of the Polish Academy of Sciences and the representative of the Committee on Acoustics of the Polish Academy of Science at the International Institute of Noise Control Engineering (I-INCE).

Contributors

Witold Mikulski, BEng, PhD, DSc, is an assistant professor at the Department of Vibroacoustic Hazards of the Central Institute for Labour Protection—National Research Institute. In 2002, he defended his doctoral thesis on architectural and urban acoustics concerning the emission and propagation of acoustic energy radiated by large surface sources. He has been a doctor of science since 2020. In his professional work, he is engaged in the topics of noise reduction (including ultrasonic noise) in the work environment and room acoustics (open-plan offices and rooms for verbal presentations in particular). He is the author/co-author of more than 65 peer-reviewed scientific publications and about the same number of scientific reports presented at international and domestic conferences. He is a certified lecturer, giving classes at higher education institutions, postgraduate studies, and training programs on noise in the work environment. He was a supervisor for more than 20 engineer's and postgraduate theses.

Leszek Morzyński, BEng, PhD, is a deputy head at Department of Vibroacoustic Hazards of the Central Institute for Labour Protection—National Research Institute. He is involved in issues related to active noise reduction. In 2003, he defended his doctoral thesis on the application of neural networks in active noise reduction systems. Other areas of his scientific activity include the processing of acoustic signals, electronic device engineering (including wireless sensor networks) used in noise-related hazard reduction, application of optimization methods in acoustic hazards control, noise measuring methods, and surround sound. He is the author or co-author of 24 peer-reviewed scientific publications, including 3 monographs and 7 patented technical solutions.

Jan Radosz, BEng, PhD, is a head of the Laboratory of Noise at the Department of Vibroacoustic Hazards of the Central Institute for Labour Protection—National Research Institute. In 2016, he defended his doctoral thesis on the acoustic quality of rooms used for verbal communication. In his professional work, he specializes in the fields of noise reduction in the work environment, room acoustics, and sound measurement methods. He is the author/co-author of 29 peer-reviewed scientific publications. He is also a lecturer at the CIOP-PIB Centre for Education where he gives lectures on noise control in the work environment.

Grzegorz Szczepański, BEng, MSc, is an engineer at the Laboratory of Active Methods of Noise Reduction at the Department of Vibroacoustic Hazards of the Central Institute for Labour Protection—National Research Institute. He participated in 14 research projects in total (in 4 as the main contractor), including a project under the Horizon 2020 framework program. He is the author/co-author of 14 scientific and conference publications. His research interests include noise visualization methods and mechanical engineering related to measurement processes automation.

Series Editor

 **Professor Danuta Koradecka**, PhD, DMedSc, director of the Central Institute for Labour Protection —National Research Institute (CIOP-PIB), is a specialist in occupational health. Her research interests include the human health effects of hand-transmitted vibration; ergonomics research on the human body's response to the combined effects of vibration, noise, low temperature and static load; assessment of static and dynamic physical load; development of hygienic standards as well as development and implementation of ergonomic solutions to improve working conditions in accordance with the International Labour Organisation (ILO) conventions and European Union (EU) directives. As an author, she has more than 200 scientific publications and several books published on occupational safety and health in her repository.

***

The "Occupational Safety, Health, and Ergonomics: Theory and Practice" series of monographs is focused on the challenges of the twenty-first century in this area of knowledge. These challenges address diverse risks in the work environment emerging from chemical (including carcinogens, mutagens, endocrine agents), biological (bacteria, viruses), physical (noise, electromagnetic radiation), and psychophysical (stress) hazards. Humans have been in contact with all these risks for thousands of years. Initially, the intensity of these risks was lower, but over time it has gradually increased, and now too often exceeds the limits of man's ability to adapt. Moreover, hazards to human safety and health, so far assigned to the work environment, are now also increasingly emerging in the living environment. With the globalization of production and merging of labour markets, the practical use of knowledge on occupational safety, health, and ergonomics should be comparable between countries. The presented series will contribute to this process.

The Central Institute for Labour Protection—National Research Institute, conducting research in the discipline of environmental engineering, in the area of the work environment and implementing its results, has summarized the achievements— including its own—in this field from 2011 to 2019. Such work would not be possible without cooperation with scientists from other Polish and foreign institutions as authors or reviewers of this series. I would like to express my gratitude to all of them for their work.

It would not be feasible to publish this series without the professionalism of the specialists from the Publishing Division, the Centre for Scientific Information and Documentation, and the International Cooperation Division of our Institute. The challenge was also the editorial compilation of the series and ensuring the efficiency of this publishing process, for which I would like to thank the entire editorial team of CRC Press—Taylor & Francis.

***

This monograph, published in 2020, has been based on the results of a research task carried out within the scope of the second to fourth stage of the Polish National Programme "Improvement of safety and working conditions" partly supported—within the scope of research and development—by the Ministry of Science and Higher Education/National Centre for Research and Development, and within the scope of state services—by the Ministry of Family, Labour and Social Policy. The Central Institute for Labour Protection—National Research Institute is the Programme's main coordinator and contractor.

1 Basic Concepts and Quantities Characterizing Sound

Dariusz Pleban

From the physical point of view, *sounds* (acoustic vibrations) are mechanical vibrations of an elastic medium (gas, liquid, or solid state matter). Such mechanical vibrations may be considered as an oscillatory motion of particles of the medium relative to their equilibrium positions, resulting in variations of local pressure in the medium relative to the static pressure (atmospheric pressure in the case of airborne sounds). Such changes in pressure (or disturbances of the equilibrium of the medium) propagate in the form of a sequence of local concentrations and expansions of medium particles in the space surrounding the source of the vibrations creating a *sound wave*. The difference between the instantaneous value of pressure in a medium in the course of the passage of sound waves and the static pressure value is called the *sound pressure* or the *acoustic pressure*, denoted p, and expressed in pascals.

Assuming that the sound pressure (as well as all the quantities characterizing the sound wave) vary harmonically (sinusoidally) as functions of both time *t* and the position of an observer *x*, one deals with the simplest form of wave motion in which particles of the medium perform simple harmonic motion and such disturbances propagate uniformly along a straight line. As a result, the *harmonic sound wave* that occurs is described with the following formula:

$$p(x,t) = A_p \sin\left[2\pi\left(t/T + x/\lambda\right) + \Phi\right] \tag{1.1}$$

where:

$p(x, t)$ (Pa): instantaneous sound pressure

A_p (Pa): amplitude of sound pressure variations

t (s): time

T (s): period of vibration

λ (m): length of the wave (called also the *wavelength*)

Φ (rad): the *phase*, a constant depending on the choice of point $x = 0$ (or the time $t = 0$)

Among the basic quantities characterizing acoustic phenomena, the following are primary:

- The *frequency f* characterizing oscillatory periodic phenomena (such as the harmonic wave), representing what is popularly called the *pitch* of sound, defined as the inverse of the period of vibration T:

$$f = 1/T \text{ Hz} \tag{1.2}$$

- The *speed of sound* (sound wave propagation velocity) c, or the velocity at which any disturbance of equilibrium propagates in the medium, is defined as the ratio of path traveled by the disturbance in an elementary interval of time to the interval value. The speed of sound can be calculated as:

$$c = \lambda / T = \lambda \cdot f \text{ m/s} \tag{1.3}$$

  The value of the speed of sound in air at temperature 20°C and under normal pressure is about 340 m/s.
- The *particle velocity u* is the speed at which a particle of medium moves in the course of the passage of a sound wave expressed in m/s and in general, is a vector quantity. Absolute value of the particle velocity (length of the particle velocity vector) is small and does not exceed a fraction of one meter per second.

When analyzing propagation of a sound wave, it is necessary to take into consideration acoustic properties of the medium. By way of analogy to optics (where properties of electromagnetic waves are considered), in the propagation of a sound wave in an inhomogeneous medium, especially when a wave encounters a boundary between two media with different acoustic properties, such phenomena as reflection, absorption, and transmission (penetration) of sound waves occur as a result of impact with a partition (e.g., a wall of an enclosure). Properties of the medium most commonly used in acoustics include:

- The *medium density ρ* (expressed in kg/m^3) in an sound wave varies together with varying sound pressure, but the variations are still very small compared to the static medium density (value of which is 1.225 kg/m^3 approximately in the so-called "*International Standard Atmosphere*", i.e., per the Standard ISO 2533:1975).
- The *acoustic impedance Z* defined as the ratio of the sound pressure to the particle velocity u specific for given medium:

$$Z = p/u \text{ Pa} \cdot \text{s/m} \tag{1.4}$$

A special case of this quantity is the *characteristic impedance Z_0* of the gaseous medium which, according to EN ISO 10534-1:2001, is defined as the impedance in a sound field per unit area of surface perpendicular to the direction of propagation of a single plane wave. Impedance Z_0 can be calculated as the product of static density of the gas medium ρ_0, and the speed of sound c:

$$Z_0 = \rho_0 \cdot c \ \mathrm{Pa \cdot s/m} \qquad (1.5)$$

Detailed descriptions of wave phenomena and principles of wave motion and harmonic motion of a particle—as well as definitions of quantities characterizing the sound-related phenomena and specific rules governing them—are discussed in numerous textbooks and monographs (e.g., [Manik 2017]). Therefore, the review of concepts and quantities of acoustics is limited to those the reader will come across when studying the present monograph.

The sound pressure values observed in the environment are characterized by a very large range of variability. For example, the sound pressure of the trilling of birds is of the order of 100 µPa, whereas the noise generated at the distance of about 500 m from a jet plane taking off corresponds to the sound pressure of about 20,000,000 µPa. On the other hand, from auditory tests, it is known that the human ear reacts to variations of sound pressure (measured in the immediate vicinity of ear) in the range from 20 to 200,000,000 µPa. The first of the values (i.e., $2 \cdot 10^{-5}$ Pa) corresponds to the *perception threshold* for a sound with frequency of 1000 Hz, whereas the second value (i.e., 200 Pa) represents the so-called *threshold of pain* (instead of the sensation of sound, pain is felt in the ear). The extremely large span of values of sound pressure makes direct use of that measurement in everyday life inconvenient. For that reason, a measure representing the sound pressure in logarithmic scale is used in practice. That measure is called the *sound pressure level*, denoted L_p, and calculated according to the formula:

$$L_p = 10 \log_{10} \frac{p^2}{p_{\mathrm{ref}}^2} \ \mathrm{dB} \qquad (1.6)$$

where:
  p (Pa): root mean square (rms) sound pressure value
  p_{ref} (Pa): the *reference sound pressure* which is often considered as the threshold of human hearing and equals 20 µPa

Simple harmonic (or sinusoidal) waves (described by Equation 1.1) are not a commonly encountered phenomenon. Usually, acoustic vibrations and sounds occurring in the environment are of a much more complex nature. In such cases, any composite periodic vibration represents a specific cyclic variability of the sound pressure in space and time. To describe the effect induced by such non-sinusoidal waveforms, the *sound pressure effective (root-mean-square, rms) value* and the *sound pressure peak value* are the quantities used in the practice of noise measurement. Therefore, it can be stated that the sound pressure level represents the sound pressure rms value level corresponding to the energy contained in the whole measured range of frequencies.

Taking a single measurement of the sound pressure level is sufficient to assess the actual intensity of a compound sound wave only in general terms (provisionally). For this reason, to determine uniquely the characteristic features of a sound waveform, its *spectrum* is determined. In other words, the content of individual harmonic components in a given acoustic vibration is established by means of measurements or calculations as a function of frequency. The technique is based on the theory according to which any periodic function can be expanded into a series of harmonic functions. Therefore, any composite sound wave can be decomposed into a sum of simple harmonic vibrations. This process is called determination of the vibration spectrum or *spectral analysis*.

As has been already mentioned, spectral analysis is carried out by determining the sound pressure levels successively in individual narrower frequency ranges, typically in octave and fractional-octave bands. The *octave* is a range or band of frequencies in which the lower limit f_1 and the upper limit f_2 satisfy the formula:

$$\frac{f_2}{f_1} = 2 \qquad (1.7)$$

Any octave can be divided into three subsequent frequency bands called *one-third octave bands*. In case of the one-third octave band, the ratio of its limit frequencies is:

$$\frac{f_2}{f_1} = \sqrt[3]{2} \qquad (1.8)$$

Individual frequency bands are characterized by the *mid-band frequency f_i*, which is determined from the formula:

$$f_i = \sqrt{f_1 f_2} \ \text{Hz} \qquad (1.9)$$

Usually, it is assumed that acoustic vibrations audible to humans cover the frequency range from 16 Hz to 16 kHz. For that reason, with regard to the frequency, acoustic vibrations are classified as follows:

- *infrasounds*—acoustic vibrations with frequencies lower than 16 Hz,
- (audible) *sounds*—acoustic vibrations with frequencies falling into the range from 16 Hz to 16 kHz,
- *ultrasounds*—acoustic vibrations with frequencies falling into the range from 16 kHz to 1 GHz, and
- *hypersounds*—acoustic vibrations with frequencies higher than 1 GHz.

As was already mentioned in the preface to this monograph, noise is any undesired, unpleasant, annoying, onerous, or harmful sound to which the organ of hearing, other senses, and parts of human body are exposed. Typically, noise is characterized by a wide range of frequencies. In some cases, noise may be dominated by acoustic vibrations with audible, infrasonic, or ultrasonic frequencies. In view of the above, three types of noise can be distinguished:

- *infrasonic noise* defined as noise the spectrum of which includes components with infrasonic frequencies from 1 to 20 Hz (the *term low-frequency noise* is also used which encompasses the frequency range from 10 to 250 Hz);
- *noise* (that is, the audible noise by default) encompassing audible sounds frequencies included in the range from 16 Hz to 16 kHz; and
- *ultrasonic noise* defined as noise the spectrum of which includes components with high audible and low ultrasonic frequencies (from 10 to 40 kHz).

In the case of defining ultrasonic noise, it should be noted that currently there is no commonly accepted definition for noise of that type in either international standardization documents or internationally ratified legal instruments. The definition of ultrasonic noise used here covers not only the airborne ultrasounds but also sounds with high audible frequencies.

The sound pressure defined by Equation 1.1 is the fundamental quantity adopted commonly to characterize the strength or intensity of sound waves. The organ of hearing responds differently to sounds characterized by the same sound pressure amplitude and level but differing in frequency. This dependence of the subjective sensation of sound intensity (loudness) on the frequency of sounds was taken into account when developing experimentally the so-called *loudness curves* corresponding to different *frequency weighting characteristics* denoted as A, C, and Z (characteristics B and D have fallen into disuse). The *A-weighting* reflects the sensitivity of the human ear, whereas *C-weighting* follows the frequency sensitivity of the ear at much higher noise levels and is used to take peak level measurements. *Z-weighting* is just a flat frequency response within the range of audible frequencies. The corresponding characteristics are implemented in the form of suitable filters integrated into sound level meters commonly used for the purpose of noise assessment. Both filters and sound level meters must meet the requirements of the standard IEC/EN 61672-1:2013. Table 1.1 is a list of A, C, and Z frequency correction characteristic values.

The level of sound pressure corrected with the frequency characteristic A in a sound level meter is commonly called the *A-weighted sound pressure level* and when corrected with the characteristic C—the *C-weighted peak sound pressure level*. The values obtained from measurements taken with the use of sound level meters equipped with A-weighing or C-weighing filters are usually quoted in units known as *decibels A* (dBA or dB(A)) or *decibels C* (dBC or dB(C)), respectively. The maximum A-weighted sound pressure level occurring in the course of

---

TABLE 1.1

Values of A, C, and Z Frequency Correction Characteristics in dB as per IEC/EN 61672-1:2013

Frequency (Hz)	63	125	250	500	1k	2k	4k	8k	16k
A-weighting (dB)	−26.2	−16.1	−8.6	−3.2	0	+1.2	+1.0	−1.1	−6.6
C-weighting (dB)	−0.8	−0.2	0	0	0	−0.2	−0.8	−3.0	−8.5
Z-weighting (dB)	0	0	0	0	0	0	0	0	0

observation and the C-weighted peak sound pressure level (i.e., the level corresponding to the maximum instantaneous value of sound pressure corrected with the frequency characteristic C) occurring in the course of observation are—together with the so-called A-weighted equivalent sound pressure level—the fundamental quantities used to formulate hearing protection criteria in the work environment. In fact, the values of these quantities are decisive for identification of potential development of hearing damage.

The *A-weighted equivalent continuous sound pressure level* $L_{p,A,eq,T}$, a quantity used to characterize a noise varying in time or changeable exposure to noise, is defined as the average value of the A-weighted sound pressure level varying in time during a stated time interval T. The quantity is determined by the formula:

$$L_{p,A,eq,T} = 10\log_{10}\left[\frac{1}{T}\int_0^T\left(\frac{p_A(t)}{p_{ref}}\right)^2 dt\right] dB \qquad (1.10)$$

where:
  $p_A(t)$ (Pa): instantaneous value of the A-weighted sound pressure

The A-weighted equivalent sound pressure level, determined for the time of exposure to noise equaling the standardized working time (i.e., for 8-hour working day or for a working week), is called the *A-weighted noise exposure level normalized to an 8-hour working day* $L_{EX,8\,h}$ (expressed in decibels) or the *A-weighted noise exposure level normalized to an average weekly working time*, as set out in the applicable labor code, $L_{EX,w}$ and is defined by the formulas:

$$L_{EX,8\,h} = L_{p,A,eq,T_e} + 10\log_{10}\frac{T_e}{T_{ref}}\,dB \qquad (1.11)$$

or

$$L_{EX,w} = 10\log_{10}\left[\frac{1}{n}\sum_{i=0}^{n} 10^{0.1\left(L_{EX,8\,h}\right)_i}\right] dB \qquad (1.12)$$

respectively, where:
  L_{p,A,eq,T_e} (dB): the A-weighted equivalent sound pressure level determined for the exposure time T_e
  T_e (s): the exposure time
  T_{ref} (s): the reference duration of 8 h (28 800 s)
  i: number of consecutive working days in a given working week
  n: number of working days in a working week (typically 5, but can be different)

One method of determination of occupational exposure to noise is the method utilizing measurements of individual acoustic events. Individual acoustic events can be characterized with the use of the *sound exposure level SEL* which (according to the standard ISO 1996-1:2016) is defined as the constant sound pressure level which has the same amount of energy in one second as the original noise event. The sound exposure level *SEL* is calculated as:

$$SEL = 10\log_{10}\left(\frac{1}{T_0}\int_{-\infty}^{\infty}\frac{p^2(t)}{p_{ref}^2}dt\right)\,dB \qquad (1.13)$$

where:
   T_0 (s): the reference duration of 1 s
   $p(t)$ (Pa): the sound pressure
   p_{ref} (Pa): the reference sound pressure of 20 µPa

The *sound exposure E_A* is another measure of exposure to sound [Berger et al. 2003]. It expresses exposure to sound in absolute physical units (i.e., in Pa²·h) and for the A-weighted sound pressure is defined by:

$$E_A = \int_0^T p^2(t)dt\ \text{Pa}^2\cdot\text{h} \qquad (1.14)$$

The noise exposure level $L_{EX,8\,h} = 85$ dB corresponds to the exposure to noise value of 1 Pa²·h.

The phenomenon of sound wave propagation in a medium involves a transmission of energy of the related disturbance. The sound wave energy is characterized by the following notions and quantities:

- the *sound power P* of a source which is a measure of the quantity of energy radiated by a source per unit of time

$$P = \frac{E}{t}\ \text{W} \qquad (1.15)$$

   where:
   E (W·s): acoustic energy emitted by the source
   t (s): time
- the *sound intensity I* and defined as the value of sound power P related to unit surface area perpendicular to the sound wave propagation direction:

$$I = \frac{P}{S}\ \text{W/m}^2 \qquad (1.16)$$

   where:
   P (W): sound power
   S (m²): surface area

Between the sound intensity I, the sound pressure p, and the particle velocity u, the following relationship can be derived for a plane wave with the use
of Equations 1.4 and 1.5:

$$I = p \cdot u = p^2/Z_0 = p^2/\rho_0 c \text{ W/m}^2 \tag{1.17}$$

As in case of sound pressure, in view of the wide range of variability of
observed sound power and sound intensity values, the logarithmic scale is
used and the following quantities are introduced

- the *sound power level L_W*:

$$L_W = 10\log_{10}\frac{P}{P_{\text{ref}}}\text{ dB} \tag{1.18}$$

where the reference sound power value $P_{\text{ref}} = 1$ pW.

- the *sound intensity level L_I*:

$$L_I = 10\log_{10}\frac{I}{I_{\text{ref}}}\text{ dB} \tag{1.19}$$

where the reference sound intensity value $I_{\text{ref}} = 1$ pW/m².

Basic acoustic quantities characterizing properties of sound-absorbing and sound-
insulating properties of partitions are: the *sound absorption coefficient α* (characterizing sound-absorbing materials) and the *sound reduction index R* (characterizing
sound-insulating materials). One commonly used sound absorption coefficient is
the *physical sound absorption coefficient*. According to EN ISO 10534-1:2001, the
physical sound absorption coefficient is defined as the ratio of sound power penetrating through the surface of a tested material sample to the sound power carried
with the plane sound wave falling perpendicularly on the sample. A quantity derived
from the coefficient is the *directional sound absorption coefficient* determined for
the angles of incidence of a plane sound wave onto a sample other than the right
angle. Another quantity characterizing construction materials and structures from
the acoustical point of view is the *sound reduction index R*. For laboratory measurements (according to EN ISO 10140-2:2010), the sound reduction index of a partition
is determined in a special set of two reverberation test rooms, and calculated from
the formula:

$$R = L_{p_1} - L_{p_2} + 10\log_{10}\left(S/A\right)\text{ dB} \tag{1.20}$$

where:
    L_{p1} (dB): average sound pressure level in the source test room
    L_{p2} (dB): average sound pressure level in the receiving test room
    S (m²): surface area of the examined partition
    A (m²): room absorption of the receiving test room

Moreover, for the purpose of the characterization of sound-insulating properties of enclosures, a quantity called the *sound power insulation* D_W is defined which (according to EN ISO 11546-1:2009) is the reduction in sound power level obtained due to an enclosure—in other words, the difference between the sound power levels from a reference sound source without enclosure and the same source provided with the enclosure. The sound power insulation is expressed in decibels.

Further, for the purpose of theory and practice of sound wave propagation in rooms and acoustic properties of rooms, the following concepts and quantities were introduced:

- The *reverberation time T* (according to ISO 3382-1:2009) is expressed in seconds and defined as the duration required for the space-averaged sound energy density in an enclosure to decrease by 60 dB after the source emission has stopped. The reverberation time can be evaluated based on a smaller dynamic rage than 60 dB and extrapolated to a decay time of 60 dB. It is then labeled accordingly: if T is derived from the time at which the decay curve first reaches 5 and 25 dB below the initial level, it is labeled T_{20}. The reverberation time can be evaluated in terms of the mid-frequencies, and then it is labeled T_{mf} which is the arithmetic average of the reverberation times in the 500 Hz, 1 and 2 kHz octave bands, or the arithmetic average of the reverberation times in the one-third octave bands from 400 Hz to 2.5 kHz. Moreover, the reverberation time can be evaluated in specific octave bands; for example, in the octave band with mid-band frequency of 2 kHz, the quantity is labeled $T_{2\,kHz}$.
- The *speech transmission index STI* (according to ISO 3382-3:2012) is defined as a physical quantity representing the transmission quality of speech with respect to intelligibility.
- The *signal to noise ratio SNR* (according to ISO 3382-1:2009) is a signal quality measure denoting the ratio of the average raw response signal level and the average noise level, the quantity being particularly relevant for speech intelligibility. The signal to noise ratio is expressed in decibels.
- The *clarity* C_{50} is a special case of the parameter *clarity*. According to ISO 3382-1:2009, the clarity C_{50} is the logarithmic early-to-late arriving sound energy ratio, where "early" means "during the first 50 ms" and "late" means "after the first 50 ms." Most practical values of the parameter range from −10 to 20 dB. The clarity C_{50} is defined by the formula:

$$C_{50} = 10\log_{10} \frac{\displaystyle\int_0^{50\,ms} p^2(t)\,dt}{\displaystyle\int_{50\,ms}^{\infty} p^2(t)\,dt}\,\mathrm{dB} \qquad (1.21)$$

where $p(t)$ is the instantaneous sound pressure of the impulse response measured at the measurement point.

- The *relative sound strength* G_{ref}, expressed in decibels, is a special case of the *strength of sound G*. The strength of sound G is the logarithmic ratio of the integrated sound power of the measured impulse response to that of the response measured at a distance of 10 m from the same sound source in free field. The strength of sound G is 0 dB when measured in free field at the distance of 10 m from the sound source. For rooms, the standardized method (covered by ISO 3382-1:2009) prescribes the use of an omnidirectional sound source and an omnidirectional microphone to measure the strength of sound G. The source is placed, for instance, on stage and the microphone is placed at successively distant listening positions. If the measuring system is not calibrated, relative sound strength G_{ref} will be calculated instead of the strength of sound G, representing the relative sound pressure level using an excitation signal with a flat frequency spectrum at the power amplifier input. Although the relative sound strength G_{ref} contains less information than the strength of sound G, it is useful for investigation of the signal level distribution over multiple listening positions in a room. The strength of sound G is defined by the formula:

$$G = 10\log_{10} \frac{\int_0^\infty p^2(t)\,dt}{\int_0^\infty p_{10}^2(t)\,dt} \, \text{dB} \tag{1.22}$$

where:

$p(t)$ (Pa): the instantaneous sound pressure of the impulse response measured at the measurement point

$p_{10}(t)$ (Pa): the instantaneous sound pressure of the impulse response measured at a distance of 10 m in a free field

- The *A-weighted teacher's voice sound pressure level* $L_{p,A,V,1\,m}$ (according to EN ISO 9921:2003) is defined as the A-weighted sound pressure level of voice at the distance of 1 m from speaker's mouth and is expressed in decibels.
- The *A-weighted speech sound pressure level at a distance of 4 m* $L_{p,A,S,4\,m}$ (according to ISO 3382-3:2012) is defined as the nominal A-weighted sound pressure level of normal speech at the distance of 4 m from the sound source and is expressed in decibels.
- The *speech spatial decay rate* $D_{2,S}$ (according to ISO 3382-3:2012) is defined as the rate of spatial decay of A-weighted sound pressure level of speech per distance doubling and is expressed in decibels.
- The *distraction distance* r_D (according to ISO 3382-3:2012) is defined as the distance from the speaker where the speech transmission index falls below 0.50 and is expressed in meters.
- The *privacy distance* r_P (according to ISO 3382-3:2012) is defined as the distance from the speaker where the speech transmission index falls below 0.20 and is expressed in meters.

- The *room constant R* identifies the effective absorption in a space in terms of square meters of totally absorptive surface [Eargle 2002]. The room constant *R* is defined by the formula:

$$R = S\alpha / (1-\alpha) \, \text{m}^2 \qquad (1.23)$$

where:
S (m²): the room boundary surface area
α (—): the average absorption coefficient of the room
- The *equivalent absorption area A* of a surface or of an object, expressed in square meters, is the area of a surface having a sound power absorption coefficient of unity that would absorb the same amount of sound power in a reverberation room with a diffuse sound field as the object or the surface. In the case of a surface, the equivalent absorption area is the product of the area of the surface and its sound power absorption coefficient [ASA 2013].
- The *room absorption*, expressed in square meters, is the equivalent sound absorption area of the room.
- The *room absorption area standardized to 1 m² of orthographic projection of the room* $A_{1 \, \text{m}^2, \text{min\_accept}}$ is the value of room absorption area divided by the surface area of orthographic projection of the room.

REFERENCES

ASA [Acoustical Society of America]. 2013. ANSI/ASA S1.1 & S3.20 Standard Acoustical & Bioacoustical Terminology Database. https://asastandards.org/Terms/equivalent-absorption-area/ (accessed January 28, 2020).

Berger, E. H., L. H. Royster, J. D. Royster, D. P. Dricsoll, and M. Layne, eds. 2003. *The Noise Manual*. Revised 5th edition. Fairfax: American Industrial Hygiene Association.

Eargle, J. M. 2002. *Electroacoustical Reference Data*. Boston: Springer. https://link.springer.com/book/10.1007/978-1-4615-2027-6 (accessed January 28, 2020).

EN ISO 9921:2003. 2003. Ergonomics—Assessment of speech communication. Brussels, Belgium: European Committee for Standardization.

EN ISO 10140-2:2010. 2010. Acoustics—Laboratory measurement of sound insulation of building elements—Part 2: Measurement of airborne sound insulation. Brussels, Belgium: European Committee for Standardization.

EN ISO 10534-1:2001. 2001. Acoustics—Determination of sound absorption coefficient and impedance in impedance tubes—Part 1: Method using standing wave ratio. Brussels, Belgium: European Committee for Standardization.

EN ISO 11546-1:2009. 2009. Acoustics—Determination of sound insulation performances of enclosures—Part 1: Measurements under laboratory conditions (for declaration purposes). Brussels, Belgium: European Committee for Standardization.

IEC/EN 61672-1:2013. 2013. Electroacoustics—Sound level meters—Part 1: Specifications. Geneva, Switzerland: International Electrotechnical Commission.

ISO 1996-1:2016. 2016. Acoustics—Description, measurement and assessment of environmental noise—Part 1: Basic quantities and assessment procedures. Geneva, Switzerland: International Organization for Standardization.

ISO 2533:1975. 1975. Standard atmosphere. Geneva, Switzerland: International Organization for Standardization.

ISO 3382-1:2009. 2009. Acoustics—Measurement of room acoustic parameters—Part 1: Performance spaces. Geneva, Switzerland: International Organization for Standardization.

ISO 3382-3:2012. 2012. Acoustics—Measurement of room acoustic parameters—Part 3: Open plan offices. Geneva, Switzerland: International Organization for Standardization.

Manik, D. N. 2017. *Vibro-Acoustics: Fundamentals and Applications*. Boca Raton: CRC Press, Taylor & Francis Group.

2 Sound Field Visualization in Noise Hazard Control

Grzegorz Szczepański

CONTENTS

2.1 INTRODUCTION

The fundamental and most effective measure by means of which workers can be protected against noise at their workstations consists in elimination of the hazard at its source. In the case of machines, designers focus mainly on maximizing the operating potential of the product with much less significance attached to issues concerning emission of noise. For that reason, predictions of machine noise emission levels are rarely undertaken early on in the stage of numerical simulations but are rather postponed to the prototype construction phase or even to the stage of installation of already manufactured machines in the work environment.

Noise emission can be characterized by a number of physical quantities, values of which are determined with the use of normalized methods. The fundamental quantities are the A-weighted sound power level and the A-weighted sound pressure level measured at workstations and at other specified positions [Pleban 2012]. Values of these parameters characterize the emitted sound energy but do not identify individual sources of emissions. An entirely different source of information are graphs and images depicting (in the form of color maps) values of quantities characterizing the sound emitted by machines and devices at precisely defined points in the space surrounding them, known as sound field parameter visualizations.

Visualizations of sound field parameters are methods of presentation of information about the sound field allowing the depiction of effects, such as sound wave reflection and diffraction, in a graphical space of sound parameters [Weyna 2003] and/or revealing the sound energy propagation direction [Weyna 2005].

These visualizations can be used to locate individual sources emitting sound energy, determine the directivity of noise-emitting sources, and evaluate effectiveness of noise control solutions.

To create a sound field visualization, it is necessary to have a set of results of measurements of sound field parameters taken at precisely defined points in space. The parameters which are the subject of visualization are the sound pressure level and the sound intensity level. The measurements necessary to work out a visualization are carried out with the use of systems comprised of large numbers of measuring transducers; however, there are also visualization techniques in which measurements can be taken with the use of a single probe guided along a path in space around given noise source. Recording and conversion of signals acquired in the course of measurements into quantities characterizing the sound field is followed by developing a representation of the quantities in graphical form on a three-dimensional or two-dimensional image of the examined object. This enables the identification of both specific features of the sound field and acoustical properties of the object. Images created with the use of visualization techniques can be widely applied in the analysis of machine prototypes constructed as part of the design process and in the construction industry (e.g., for checking tightness of sound-insulating screens and walls of buildings). The present chapter also includes presentation of research studies (carried out with the use of up-to-date measuring systems) enabling visualization of sound field parameters.

2.2 THE STATE OF THE ART

Visualizations of sound field parameters are usually the result of mathematical calculations performed on acoustic signals recorded with the use of microphones, according to a selected method, enabling determination of sound field parameters in a selected portion of space. The two most popular techniques are the beamforming method and acoustic holography [Kim and Choi 2013]. The original application of the beamforming method was in radio engineering where the technique was used to improve directivity of radio signal transmission and reception [Athanossios 2015], but it turned out to be equally useful in identification of sound emission sources. The method requires that multiple-point sound pressure measurements be taken simultaneously. To this end, devices commonly called acoustic cameras or microphone arrays are used. These are often paired with image recording cameras which, apart from recording the sound pressure waveforms for each of the microphones, record the video image in a plane parallel to that of the microphone array. The signals obtained from individual microphones and amplitude-phase relationships between them are used to determine sound field parameters on a plane situated at a selected distance from the array. The array shape is optimized with regard to conditions in which the measurements are to be taken and the nature of the examined noise. Dimensions of the array and the number of microphones used affect detection performance and resolution with which the location of emission sources can be determined using visualization. To increase the resolution, deconvolution techniques are used [Brooks and Humphreys 2004]. The size of the examined noise source can exceed significantly the dimensions of the microphone array. Indisputable advantages of the beamforming method include its expeditiousness, as a single measurement taken with the use of a microphone array is often sufficient to obtain a sound field visualization. What is more, the method can be used in dynamically changing situations (e.g., for non-stationary noise measurements and studies on traffic generated by mobile objects). Beamforming methods turn out to be very effective in free field conditions and these are the environments in which they are typically used. Measurements carried out with the use of the acoustic camera should be taken in the far field [Johnson and Dudgeon 1993], with the distance from the examined source ranging typically from one to several wavelengths and being subject to analysis in each individual case [Christensen 2004]. Difficulties encountered in the practice are visualization disturbances when noise sources are coherent (e.g., occurrence of image sources resulting from sound wave reflections from obstacles) [Schwartz et al. 2015].

Over time, the beamforming method has been subject to numerous modifications and improvements resulting in increasingly reliable results and widening of the range of options in the scope of sound source analyses. These have found application mainly in studies on noise generated by rail vehicles, motor vehicles, and aircraft. An interesting example of the application of the beamforming technique is reported in [King and Bechert 1979], where the method was used for the purpose of identification of the main sources of noise generated by a train. The beamforming method was also used to locate sound sources at the wheel-rail contact point [Barsikow et al. 1987]. In parallel, studies were carried out on the possibility of the use of two-dimensional microphone arrays for the same purpose [Takano et al. 1992] and the effect of array type on the precision with which location of noise sources generated by rail vehicles can be determined [Brühl and Schmitz 1993; Barsikow et al. 1996; Kitagawa and Thompson 2010]. The beamforming method was also used to predict aerodynamic noise generated by high-speed (above 300 km/h) rail vehicles [He and Jin 2014; Thompson et al. 2015]. In the automotive industry, repeated attempts were made to optimize the number of microphones used in measurements of noise due to moving automobiles [Kook et al. 1999]. Other studies pertained to improvement of identification resolution and elimination of the Doppler effect [Bolton et al. 2000; Michel et al. 2004; Yang et al. 2011] and examination of the sound pressure level characterizing individual components of noise produced by a moving vehicle [Ballesteros et al. 2015]. In studies oriented toward the aircraft industry, the beamforming method was used to identify aircraft components responsible for noise emission in the course of takeoff, landing, and flight at cruising altitude [Sijtsama 2010; Simons et al. 2015; Merino-Martinez et al. 2016a]. As in automotive industry studies, the quest continued for methods to eliminate the Doppler effect [Howell et al. 1986] and ways to improve resolution of the identification process, especially in the low frequency range [Merino-Martinez et al. 2016b]. Besides these examples of research on road and aviation transport noise, the beamforming method was widely used in studies on noise generated by wind turbine farms [Oerlemans et al. 2007; Dougherty and Walker 2009; Yang 2013; Pannert and Maier 2014; Buck et al. 2016] and identification of industrial noise sources [Bai and Lee 1998; Allozola and Vonrhein 2019]. Applicability of the beamforming method in researching noise in working environments may be found in measurements of noise emission from machines installed outdoors, including identification of noise sources in specialized vehicles used in the construction and mining industries [Yantek et al. 2007].

Acoustic holography is a method offering the possibility of obtaining a complete description of a sound field in three-dimensional space [Maynard et al. 1985]. As opposed to the beamforming method, acoustic holography measurements are taken with the use of two-dimensional microphone arrays or special acoustic transducers enabling direct measurement of the vector component of the sound field—the particle velocity (known also as the acoustic particle velocity). The measurements are typically carried out in the sound nearfield, but taking measurements in the farfield is also possible. The result of application of acoustic holography may not only create a sound pressure level distribution image but also a depiction of particle velocity and sound intensity vector fields in three-dimensional space [Williams 1999; Hayek 2008].

Information obtained with the use of acoustic holography, apart from enabling the location of a particular spot responsible for the highest noise emission levels on an examined source (which is the main domain for application of beamforming methods), also offers insight into the way sound waves, generated as a result of mechanical vibration of an examined object, propagate in a medium. This enables creation of three-dimensional maps representing graphically energy distribution and vector processes occurring in the sound field. As opposed to the beamforming method, microphone arrays used in acoustic holography must have a regular shape and are typically set up using a rectangular layout.

To date, numerous variations of the acoustic holography method have been developed, including STSF (Spatial Transformation of Sound Fields), SONAH (Statistically Optimized Near-field Acoustic Holography), and WBH (Wideband Holography), which are used in many industries to identify noise emission sources. In the automotive industry, the dominant subject of research is noise emitted from the automobile engine compartment [Hald 2001; Lafon and Antoni 2008; Lu and Jen 2010] and individual vehicle mechanisms such as gearboxes [Derouiche et al. 2012]. Another area of application is machine fault diagnostics based on data acquired with the use of acoustic holography methods [Hou et al. 2011; Lu et al. 2013]. In the aircraft industry, acoustic holography is employed in examination of noise sources inside airplane fuselages [Williams et al. 2000; Sklanka et al. 2006], in research on the mechanism of noise generation in jet engines [Lee and Bolton 2007; Wall et al. 2014, 2016], and in examination of aerodynamic noise [Lee et al. 2003; Chelliah and Raman 2014]. In the shipbuilding industry, acoustic holography methods are used to detect sound sources inside ship hulls [de la Croix et al. 1997] and test tightness of ship bulkheads [Weyna 1995]. Acoustic holography is also used in diagnostics of the technical condition of machines and devices. In that area of application, acoustic holography methods are used in tests of compressors [Mann and Pascal 1992; Cho 2015], electric motors [Orman and Pinto 2013; Bonanomi 2017], and various household appliances, including washing machines [Chyariottia et al. 2010]. Literature on the subject includes reports concerning the use of the acoustic holography technique in acoustical assessment of medical equipment [Kreider et al. 2013; Sapozhnikov et al. 2015] and in tests concerning tightness of buildings [Chelliah et al. 2015; Patel et al. 2015]. Acoustic holography methods have shown high potential in research on noise emissions generated by machines in the work environment [Peterson et al. 2012; Attendau and Ross 2016]. These methods can be used as elements of the design process and prototype testing of machines and also in development and testing of new hearing protection devices for workers (collective protection measures).

The contemporary market of sound field visualization apparatus for science and research sees a trend toward offering upgraded microphone arrays dedicated to beamforming and acoustic holography. However, other techniques are also capable of providing information on noise emitted by machines installed at workstations. Visualizations of sound pressure level distribution patterns can be created based on results of microphone measurements taken at a plurality of points in space surrounding the examined noise source.

As an example of such a solution, it is worth noting the method of sound pressure level (SPL) distribution visualization in selected planes with the utilization of a laboratory microphone measuring system developed in the Central Institute for Labour Protection–National Research Institute (CIOP-PIB). The system is installed in the Institute's acoustic chamber offering conditions close to those characterizing the free field and comprises 80 microphones of Audix TM-1 type distributed evenly on chamber walls and suspended at the height of about 2 m under the ceiling forming an ellipse in central portion of the room where examined sources are placed for examination. The arrangement of microphones can be modified depending on current needs and overall dimensions of the examined noise source. The basic arrangement of measuring microphones in the laboratory is shown in Figure 2.1.

The system microphones are connected to analog-to-digital (A/D) AxC-Ax4M converter cards installed in a dedicated docking panel equipped with a card compatible with Dante (Digital Audio Network Through Ethernet) standard. The docking

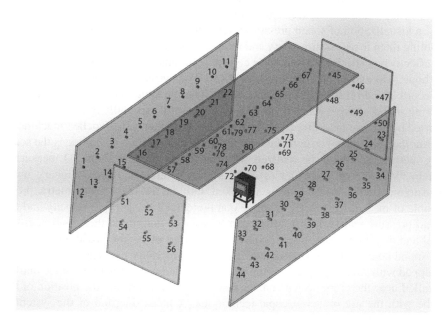

FIGURE 2.1 Arrangement of microphones in the measuring system installed in the acoustic chamber.

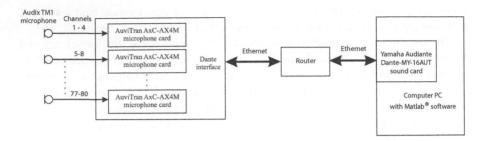

FIGURE 2.2 A schematic diagram of the measuring system installed in the acoustic chamber.

panel is connected via Ethernet with a sound card installed in a PC-class computer. The Dante interface-based Ethernet connection provides transmission of measurement signals from A/D converters to the sound card. Input channel parameters (especially the gain) of the used converter cards can be software-modified from the system configuration panel. Figure 2.2 shows a schematic diagram of the measuring system installed in the acoustic chamber.

Signals from measuring microphones are recorded and their effective (root-mean-square) values are determined with the use of the MATLAB® Simulink package. When ASIO controllers are used, signals are acquired in digital form from the sound card and sent to the MATLAB software environment where they can be arbitrarily processed. A multi-microphone measuring system is used to measure sound pressure levels on planes parallel to the walls of the acoustic chamber and at selected points in the central portion of its space. This allows researchers to obtain, in a relatively easy way, a visualization of sound pressure level values generated by quite large sources enabling them to assess directivity and intensity of the emitted noise and thus evaluate effectiveness of the adopted noise control solutions based on active noise control methods [Morzyński and Szczepański 2018]. The system also enables researchers to record sound pressure level variations in dynamic situations (e.g., in the case of moving sound sources) which turns it into a tool useful especially in studies on noise emitted by, for instance, drones, forklift trucks, and other self-propelled machines.

Another example of a sound field parameters visualization technique is the scanning method which can be realized with the use of Scan & Paint 3D measuring system (Microflown Technologies). The system is intended for stationary noise measurements. The measurements are taken with the use of an anemometric probe capable of measuring directly all three components of the particle velocity and the sound pressure value which together can be used to determine the sound intensity level and direction at any given point of space. That offers the possibility of locating sound sources and assessing their radiation directivity patterns. The system is equipped with a stereoscopic tracking camera coupled with a tracking sphere marker installed near the probe which enables researchers to determine the position of the probe with the use of stereoscopic techniques. A block diagram of the system is shown in Figure 2.3.

To generate a visualization in the form of an image, it is necessary to create a model of the examined object in the form of a solid and import it into the software

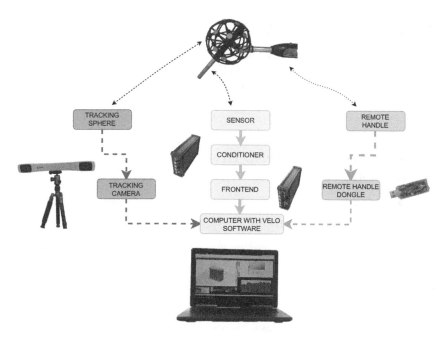

FIGURE 2.3   A block diagram of the Scan & Paint 3D system.

environment of the system visualization system. The file representing the object's geometry may be generated by a designer's CAD software or obtained directly by 3D scanning performed, for instance, with the use of a tablet equipped with optical surface scanning adapter. This option is particularly useful in industrial conditions where a solid model of a quality sufficient to take accurate measurements can be obtained readily.

Once the model is imported, the matching procedure is launched which involves selection of a number of representative points on the solid model and marking them physically on the examined object. Results of the matching procedure inform about the precision to which measurement points are ascribed to specific points in space. Next, it is possible to take the measurement by moving manually the measuring probe in the space surrounding the examined object. In the majority of cases, the probe is led along a meandering path at some distance from the object. The obtained measurement results can be used as input data for the acoustic holography method [Luo et al. 2019], visualized in the form of sound intensity vectors distributed around the model of the examined source, or shown as a projection of selected intensity component onto a plane in a selected measurement area [Serraris 2016]. In view of the fact that measurements are carried out in 3D space, the results are assigned to certain elementary volumes (of the so-called cubic lattice) defined in space in which the measurement was taken. The cubic lattice constant (the cubic cell length) is defined by the user in the course of analysis of the collected measurement results. In case of multiple measurements taken within the area of the same cubic lattice element, the measurement results are averaged.

The measuring methods described above and systems developed as embodiments of the techniques are used in studies of noise emitted by machines and devices in both laboratory and industrial site conditions (apart from the above-described microphone measuring system constructed specifically for the acoustic chamber in which it is installed). In the next section, example results of studies carried out with the use of these methods are presented together with certain lessons learned from these studies.

2.3 APPLICATION OF SOUND FIELD VISUALIZATION METHODS TO NOISE EMISSION STUDIES

2.3.1 EXAMPLES OF VISUALIZATIONS BASED ON STUDIES CARRIED OUT IN LABORATORY CONDITIONS

Conventional measurements of parameters characterizing noise generated by machines and devices, taken with the use of portable noise meters/analyzers, constitute the first step in the process of creating a sound field visualization. An image enabling one to identify noise emission sources and noise propagation paths may take various forms. The first-choice, simplest, and most popular forms are visualizations generated on the basis of measurements of parameters characterizing the sound field, such as the sound pressure level and the sound intensity level at given measurement points in space around the examined object. The measurements are taken at specific points of space (e.g., on a selected measurement surface), and values of the parameters in areas where no measurements were effected are interpolated. Another option consists in presentation of measurement results in the form of a spatial visualization. That is the most typical method of visualization used for sound field disturbance effects, sound wave vector forms, and sound wave shapes for disturbances occurring at specific frequencies.

As an example introducing to the issue of visualization, results of research carried out in laboratory conditions are presented. The subject of the study was sound radiation from a sound-insulating enclosure with a window on one side and constructed with the option of mounting single-layer or multi-layered plate baffles against the window. The source of excitation was a loudspeaker placed inside the sound-insulating enclosure. The excitation signals were tones generated with the use of Brüel & Kjær 1049 generator. The sound emitted from the enclosure was examined with the use of three different methods—the SPL distribution visualization method, the acoustic holography technique, and the scanning method.

The sound pressure level distribution pattern visualization method includes the step of recording noise emitted by the examined source with the use of a microphone measuring system, determination of sound pressure levels at specifically chosen measurement points, and the step of creating the visualization. The measurements were carried out for three configurations of the window:

- with the window open,
- with a baffle in the form of an aluminum plate with thickness of 2 mm mounted on window axis, and

- with an installed baffle comprised of two aluminum plates, each with thickness of 2 mm, one situated in the window axis, and the other fixed by means of an aluminum frame at the distance of 25 mm in such way that the area between the plates formed a closed air chamber.

Four cases of excitation were examined in the form of sine sound waves with the frequency of 70 Hz, 165 Hz, 200 Hz, and 310 Hz. To assure the same test conditions for each of the three configurations of the examined system, the excitation signal sound pressure level value measured in the open window plane was the same in each case and equaled 105 dB. A view of the sound-insulating enclosure on the measuring setup installed in the acoustic chamber is shown in Figure 2.4.

The baffles were fixed rigidly to the enclosure by means of a clamping frame and fourteen M6 bolts tightened with the moment of force of 5 Nm. First, it was necessary to measure the background noise level close to the tested machine, values of which are presented in Figure 2.5.

The highest background sound pressure level recorded in the acoustic chamber concerned low-frequency components (below 63 Hz) which should not significantly affect the results of measurements taken for excitation with the frequencies indicated above. In the test, 67 microphones were used, making up the microphone measuring system arranged on planes parallel to chamber walls and ceiling. Visualizations obtained from results of the measurements are presented in Figures 2.6 through 2.11.

These visualizations illustrate sound pressure level distribution patterns recorded on a plane parallel to and 50 cm away from acoustic chamber walls. The results are presented on two different sound pressure level scales, of which one is adjusted to results obtained in the range of a single excitation frequency applied to different window configurations (Figures 2.6 and 2.9) and the other is adapted to results obtained in the given area for all excitation signals (Figures 2.10 and 2.11). In case of tests in

FIGURE 2.4 The sound-insulating enclosure with open window on the measuring set up in the acoustic chamber.

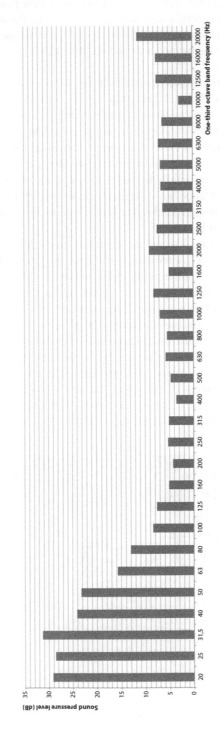

FIGURE 2.5   Results of sound background measurements at the measuring set up in the acoustic chamber in one-third octave frequency bands.

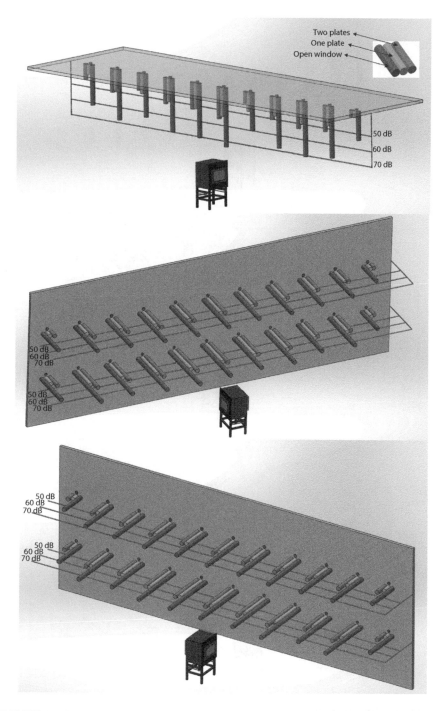

FIGURE 2.6 A visualization of the sound pressure level distribution for excitation with the frequency of 70 Hz.

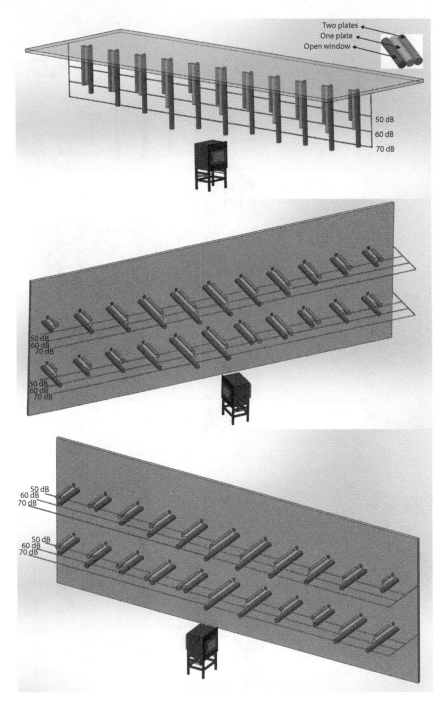

FIGURE 2.7   A visualization of the sound pressure level distribution for excitation with the frequency of 165 Hz.

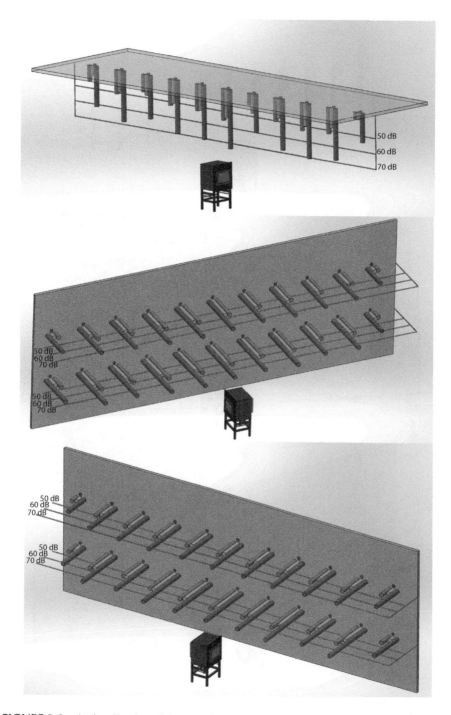

FIGURE 2.8 A visualization of the sound pressure level distribution for excitation with the frequency of 200 Hz.

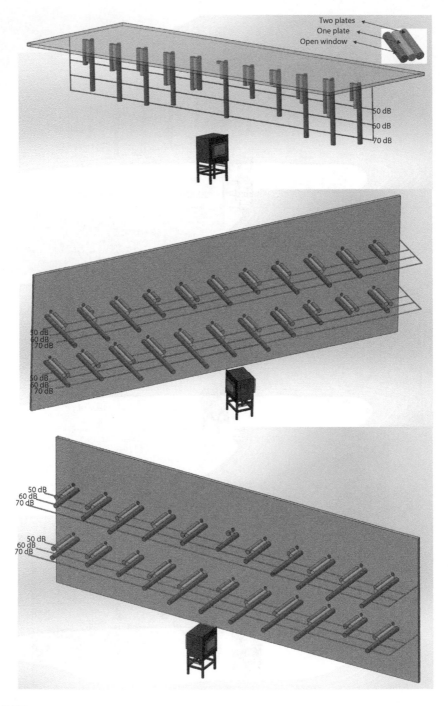

FIGURE 2.9 A visualization of the sound pressure level distribution for excitation with the frequency of 310 Hz.

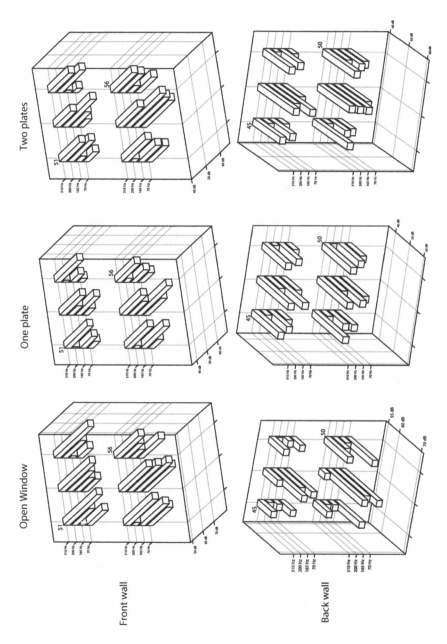

FIGURE 2.10 A visualization of the sound pressure level distribution on the front wall and the back wall of the acoustic chamber.

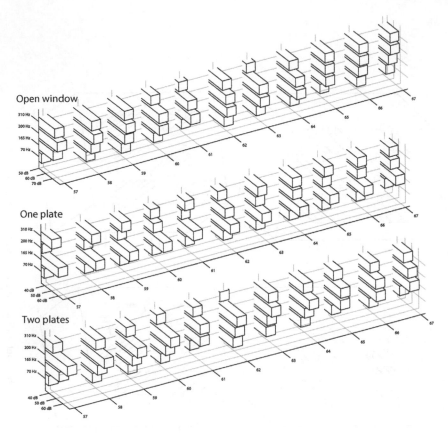

FIGURE 2.11 A visualization of the sound pressure level distribution on the ceiling of the acoustic chamber.

which the excitation signal was a tone with the frequency of 70 Hz, the highest sound pressure level values were recorded from microphones situated in lower rows of the left-hand and right-hand side planes of the chamber situated closest to the source. The same trend can be observed for all the three window configurations. Regardless of configuration of the window, the highest value of sound pressure level for this specific excitation frequency was recorded from microphone No. 16 situated on the plane parallel to the left-hand side wall of the chamber.

Visualizations of the sound pressure level for the case of excitation with the signal frequency of 165 Hz indicate that the highest sound pressure level values are observed at points corresponding to the row of microphones situated on the upper plane parallel to the chamber ceiling. High sound pressure level values were also detected by individual microphones placed on planes parallel to chamber side walls. The test results for the excitation signal with the frequency of 200 Hz are diversified. Relatively low sound pressure level values were observed on the front and the rear plane as well as at microphone location points on the left-hand and right-hand side planes situated farthest away from the source. In case of excitation with the frequency of 310 Hz, the highest sound pressure level was recorded from microphones situated in the close vicinity of the source, on the front plane. It is worth noting that installing the second baffle resulted in an increase of the sound pressure level value at some of the measurement points. The baffle, therefore, became a secondary noise source. The visualizations representing test results for a given excitation frequency depict the nature of the variation of sound field parameters depending on the examined configuration of the window, whereas the visualization corresponding to test results obtained for given arrangement of microphones (i.e., given measurement plane of the chamber) for different excitation frequencies illustrate the effect of the configuration on the directivity pattern of the sound radiated from the sound-insulating enclosure. Even without referring to specific sound pressure level values, such images may turn out to be useful for the purpose of preliminary assessment of sound radiation patterns and for identification of specific components (or specific areas on the machine) which should be subjected to detailed examination in order to locate precisely the noise source(s) (acoustic holography, scanning method). The microphone arrangement configuration for the examined case was adapted to the overall dimensions of the source. In case of noise sources with smaller dimensions it is possible to move the farthest microphones toward the central portion of the chamber in order to compact the array of microphones on the measurement surface.

Another visualization method tested in laboratory conditions was acoustic holography. Nowadays, the market offers microphone arrays of various brands which can be used when applying the acoustic holography method for the purpose of identification and characterization of noise sources. In measurements carried out in both laboratory and industrial conditions, the microphone array marketed as Paddle 2×24 AC pro Acoustic Camera (gfai tech GmbH) was used as shown in Figure 2.12.

The array is designed for taking measurements in the nearfield within the frequency range from 30 Hz to 2 kHz. A system of 48 microphones (two layers, each

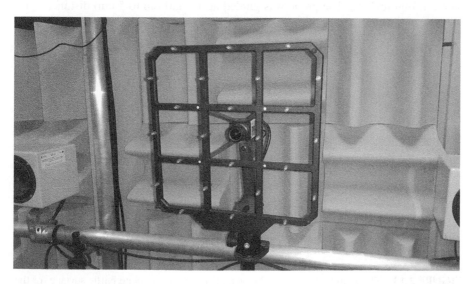

FIGURE 2.12 The microphone array used in the study.

comprising 24 microphones) is used to take sound pressure measurements in two planes. This arrangement produces data that can be utilized for visualization of sound field parameters using such methods as nearfield acoustic holography or beamforming. In the course of measurements, the array was connected to the measurement data acquisition system mcdRec_721 and a PC-class computer on which NoiseImage version 4.7.1 software was installed. The study was carried out for the case of excitation with the frequency of 200 Hz for the enclosure window configuration with a double plate baffle installed. The measurements were taken with the microphone matrix positioned parallel to and at the distance of 40 cm from the baffle plane and parallel to and at the distance of 30 cm from each of the side surfaces of the sound-insulating enclosure. Results of the measurements were analyzed with the use of the nearfield acoustic holography (NAH) method.

The sound intensity level distribution on the plate surface corresponds to the plate structure vibration distribution which enables identification of the plate regions with highest vibration amplitudes. In the visualization shown in Figure 2.13, two areas can be clearly distinguished in which high sound intensity levels occur. The maximum observed sound intensity level value was 85.7 dB. On the other hand, in visualizations presented in Figure 2.14 it is possible to identify isolated areas corresponding to radiation emitted from two sides of the sound-insulating enclosure.

The last of the tested visualization methods was the scanning method. Compared to the earlier techniques, the scanning method is very time-consuming. Results of measurements from the SPL distribution method and acoustic holography method were obtained from measurements lasting 8 s. In case of the scanning method, the measurement-taking process took 37 min and 30 s. The excitation signal was the same (with the frequency of 200 Hz) as was applied in measurements taken with the use of the microphone array. The tests were carried out in the area close to the window provided in the sound-insulating enclosure and on its left-hand side (cf. Figure 2.15). The probe was guided at a small (up to 5 cm) distance from

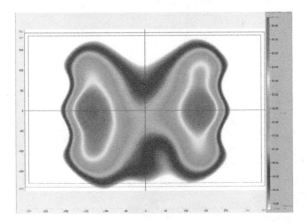

FIGURE 2.13   The sound intensity level distribution pattern on plate baffle surface for the excitation signal frequency of 200 Hz.

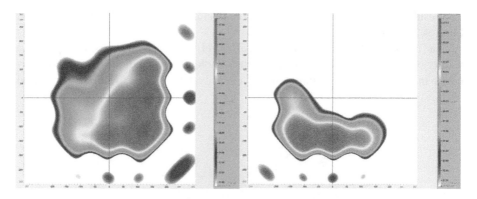

FIGURE 2.14 The sound intensity level distribution pattern on left-hand (left panel) and right-hand (right panel) side surface of the sound-insulating enclosure.

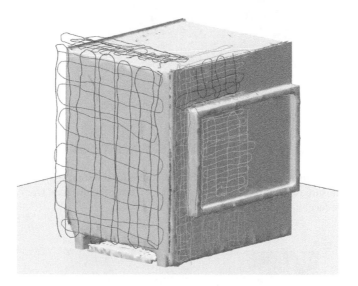

FIGURE 2.15 Noise measurement trajectories used in testing the sound-insulating enclosure with double baffle with the use of scanning method.

the plate face along meandering trajectories and on several more distant planes (about 15 and 30 cm from the plate front surface).

The visualizations were compiled for cubic lattice with the lattice constant (cube edge length) of 50 mm in which the following sound field parameters were displayed:

- the sound intensity level on a plane parallel to the plate baffle at the distance of from 10 to 60 mm from the baffle face,
- the sound intensity level on a plane perpendicular to the plate baffle tangent to its symmetry axis, and
- the sound intensity level on a plane parallel to the baffle surface outside the aluminum frame.

The results are presented in the form of sound intensity level distribution patterns in Figures 2.16 through 2.18.

The maximum recorded sound intensity level value was 87.2 dB. Compared to results obtained with the use of the acoustic holography method, there are differences in the recorded maximum sound intensity level values, whereas the sound

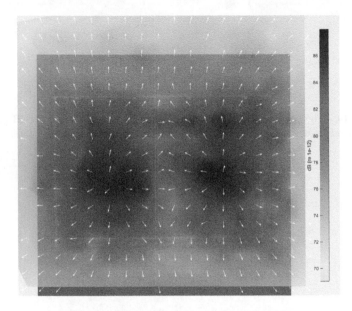

FIGURE 2.16 Sound intensity level distribution pattern for the sound emitted by the sound-insulating enclosure on a plane parallel to the plate baffle at the distance from 10 to 60 mm from the baffle face.

FIGURE 2.17 The sound intensity level distribution pattern for double plane baffle on a plane perpendicular to the baffle and tangent to its symmetry axis.

FIGURE 2.18 The sound intensity level distribution pattern observed on the plane parallel to the baffle surface on its outer edges (boundaries).

intensity level distribution pattern on the plate surface has a similar form. In both cases, two vibrating areas emitting the main portion of the acoustic energy radiated by the plate baffle are visible.

The accuracy with which the point with the highest amplitude of plate vibrations can be identified using the scanning method depends on the adopted cubic lattice of measurement points defined by the distance between nearest lattice nodes called the lattice constant. The visualization shown in Figure 2.18 also suggests the possibility of the occurrence of flanking transmission (wherein the sound passes around, over, or under the blocking plates). The maximum sound intensity level values were found on the right-hand side of the frame and equaled 83.8 dB. The value is just under 10 dB less than the maximum sound intensity level values obtained in the test performed close to the plate baffle. Figure 2.19 also shows a pseudo-3D visualization of sound intensity vectors depicting noise radiation coming from the left side, a portion of the front side (with window and plate baffle), and from the rear portion of the sound-insulating enclosure.

In the visualization, dark colors represent high sound intensity level values. Therefore, it can be seen that the main sound-emitting element is the plate baffle mounted in the window of the sound-insulating enclosure.

Carrying out studies on noise emissions generated by machines and/or their sound-insulating enclosures in laboratory conditions offers a number of benefits. It makes possible, for instance, maintenance of stable measurement conditions including absence of additional noise sources, low ambient noise level, and unchanging environmental conditions. Studies in industrial conditions are carried out typically for

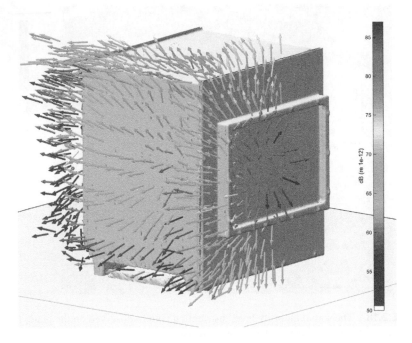

FIGURE 2.19 A pseudo-3D visualization of the sound intensity level vector distribution on surface of the left-hand side of the enclosure.

identification of noise emission sources within production lines, on machines with large overall dimensions, or on machines that require continuous supplies of additional components or raw materials to function properly. The next subsection presents examples of visualizations based on studies carried out in industrial conditions.

2.3.2 EXAMPLES OF NOISE VISUALIZATION IN INDUSTRIAL CONDITIONS

In cases involving measurements carried out under industrial conditions, the main noise sources are large industrial machines for which it is necessary to inspect the foundation location, assess the adequacy of application of the given measuring method, and select the number and location of measurement points or the survey area (in case of the scanning method). In studies carried out in laboratory conditions, the SPL distribution visualization method is a good choice when making decisions on situation of measurement points, for instance, where a microphone array should be positioned. On the other hand, in studies carried out in industrial conditions, measurements are taken using sound level meters or other systems capable of providing information about the sound pressure level existing at the measurement location together with the signal spectrum.

The first example is a presentation of results of examination of the noise emitted by an industrial furnace constituting one element in a production line for manufacture of electronic components. Although no noise levels exceeding the

FIGURE 2.20   A view of the examined source—an industrial furnace.

permissible values were recorded at workstations situated close to the furnace, the emitted noise was described by workers as bothersome. A view of the examined noise source is shown in Figure 2.20.

Three measurements were taken with the use of type 3052-A-030 input module and TEDS-class type 4191 microphone (Brüel & Kjær), one for the furnace front (door) and two of furnace sides at the distance of 0.5 m from these surfaces. The equivalent sound pressure level was determined for individual recordings (results are summarized in Table 2.1) and the frequency analysis was performed in the range from 3.15 Hz to 20 kHz in order to identify and compare positions of components dominating in the signal spectrum. The dominating components occurred in the range of frequencies up to 700 Hz, as can be seen from the spectrum of the recorded signal shown in Figure 2.21.

Since the sound pressure level value recorded on the left-hand side of the furnace was significantly higher than the values recorded at other measurement points, it was considered advisable to undertake additional sound field examination of the left side aimed at detailed identification of the noise source. From the analysis of signal spectra in Figure 2.21, it follows that dominant components were recorded for frequencies of 50 Hz, 73.5 Hz (and its harmonics 147 Hz, 220.5 Hz, 294 Hz, and 367 Hz), 171 Hz, and 392 Hz. In case of the measurement taken on the left-hand side of the

TABLE 2.1
Equivalent Sound Pressure Level for Noise Measurements Carried Out at Selected Areas Near the Industrial Furnace

Measurement Point Situation	Recording Time (s)	Equivalent Sound Pressure Level, Linear (dB)	Equivalent Sound Pressure Level, A-weighted (dB)
Furnace left-hand side	30	92.2	86.7
Furnace front	30	84.9	75.3
Furnace right-hand side	30	81.6	68.3

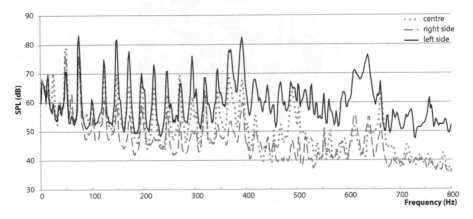

FIGURE 2.21 Spectra of signals recorded on front, left-hand, and right-hand side of the furnace.

furnace, a 658 Hz component is also visible. Next, measurements aimed at detection of noise emission sources with the use of the scanning method were continued for two areas (the first area encompassing left side of the furnace, the second covering fragments of the right side and of the furnace front). For the purpose of application of the scanning method, solid models were developed for selected furnace portions which, together with probe tracing paths, are shown in Figure 2.22.

As part of examination of the first area, measurements were taken lasting a total of 30 min, whereas the measuring time for the second area was 12 min. For visualizations of the sound field in the first area, the cubic lattice constant of 40 mm was adopted and a diversified scale of the visualized sound intensity level range (maximum and minimum values of scales in which individual components are presented are different). The visualization results are shown in Figures 2.23 through 2.27.

The highest sound intensity levels were recorded for frequencies of 50 Hz, 147 Hz, and 392 Hz. The sound intensity vectors for the frequency of 50 Hz point uniformly in a direction outward from the furnace housing which indicates that the whole furnace emits noise in the spectrum where this frequency dominates. At frequencies

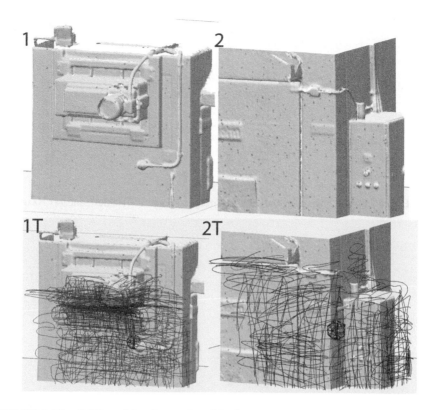

FIGURE 2.22   Solid models of areas 1 and 2 of an industrial furnace together with probe guiding trajectories used for respective measurements (1T and 2T).

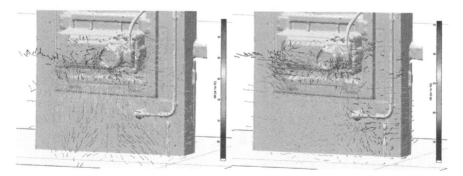

FIGURE 2.23   Sound intensity visualization for frequencies of 50 Hz (left) and 73.5 Hz (right).

of 220 Hz, 294 Hz, 367 Hz, and 392 Hz, intensity vectors point at elements of the furnace housing and an electric motor housing (together with its mount) as the main sound sources. One unambiguously identified source of noise emission for the frequency of 658 Hz was a motor element situated behind air vent grilles on the left-hand side. To confirm correctness of the obtained sound intensity distribution

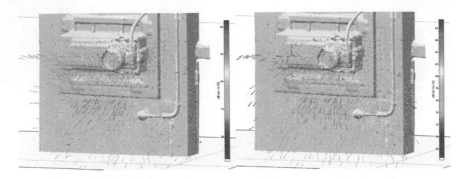

FIGURE 2.24   Sound intensity visualization for frequencies of 147 Hz (left) and 171 Hz (right).

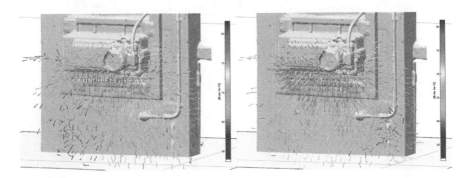

FIGURE 2.25   Sound intensity visualization for frequencies of 220 Hz (left) and 294 Hz (right).

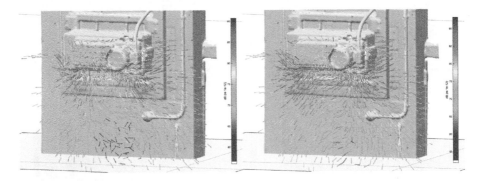

FIGURE 2.26   Sound intensity visualization for frequencies of 367 Hz (left) and 392 Hz (right).

patterns within the area of furnace elements, additional visualizations of the sound intensity level for selected measurement planes were compiled. The visualizations are shown in Figures 2.28 and 2.29.

When creating the visualization of the second area, the cubic lattice constant of 50 mm was adopted and a fixed range was used from 55 to 90 dB for presentation

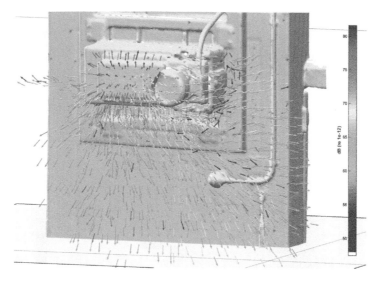

FIGURE 2.27   Sound intensity visualization for the frequency of 658 Hz.

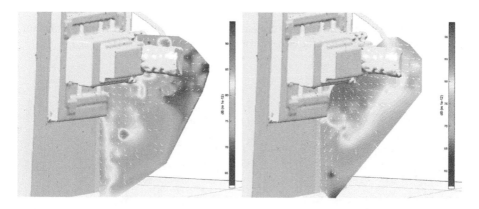

FIGURE 2.28   Sound intensity visualizations in plane perpendicular to the furnace side surface and tangent to the motor symmetry axis for 50 Hz (left) and 392 Hz (right).

of the sound intensity levels. The results for a number of selected frequencies are presented in Figures 2.30 through 2.34.

In the second area, the highest sound intensity level values were observed for the frequency of 73.5 Hz. The presented visualizations indicate that the noise emission source was the area of a joint between the control pulpit and the furnace housing. As was the case in the examination of the first measurement area, the intensity vectors for the frequency of 50 Hz are distributed evenly, therefore it should be acknowledged that all the furnace housing elements (encompassed within the adopted measurement area) generate noise for that frequency component. At frequencies

FIGURE 2.29 Sound intensity visualization in plane perpendicular to the furnace side surface and crossing the center of the element with air vent grilles for 658 Hz.

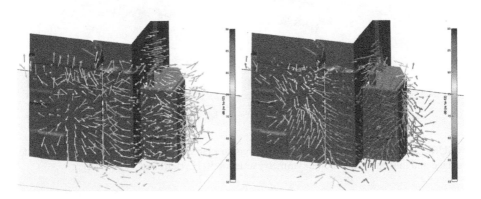

FIGURE 2.30 Sound intensity visualizations for frequencies of 50 Hz (left) and 73.5 Hz (right).

of 220 Hz, 294 Hz, 367 Hz, and 392 Hz, intensity vectors point at a hot air extractor hood installed above the furnace door identifying it as the main noise source (Figure 2.35).

In case of the frequency 658 Hz, the level of 55 dB was exceeded only at a few measurement points which indicates that the source of noise for that frequency

FIGURE 2.31   Sound intensity visualizations for frequencies of 147 Hz (left) and 171 Hz (right).

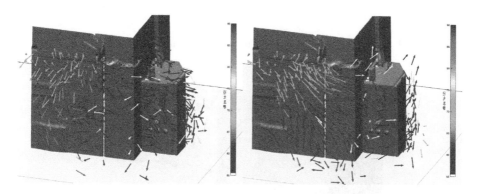

FIGURE 2.32   Sound intensity visualizations for frequencies of 220 Hz (left) and 294 Hz (right).

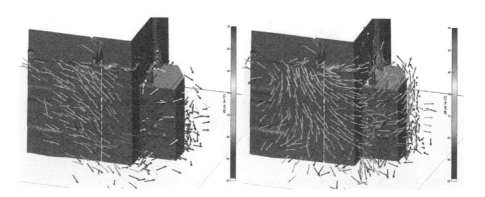

FIGURE 2.33   Sound intensity visualizations for frequencies of 367 Hz (left) and 392 Hz (right).

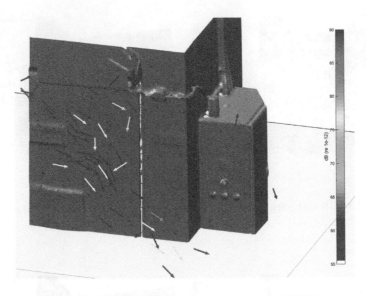

FIGURE 2.34   Sound intensity visualization for the frequency of 658 Hz.

FIGURE 2.35   A fragment of hot air extractor installed above furnace door.

occurs in the first area, the observation being consistent with earlier visualizations. The advantage of visualizations (created with the scanning method) which depict the vector parameters characterizing the sound field (the sound intensity) over visualizations which depict only the scalar component of the field (the sound pressure level) is the ability of the scanning to represent the direction and sense of the sound energy flux vector generated by the examined machines. However, the main limitation of the scanning method is the nature of the noise emitted by a specific machine. In case of impulse noise, the method is ineffective. The next example concerns a study on emission of noise occurring on an industrial shop floor near a workstation comprised of a press for drawing sheet steel articles. The examination was carried out with the use of the beamforming method by means of a microphone array. Results of the measurement were recorded within 8 s. Then fragments of the recording corresponding to individual sound events were identified and marked. Figure 2.36 shows the waveform of the signal recorded by means of the microphone array.

The visualization encompassed a wide range of frequencies (from 40 to 400 Hz) for individual sound events (i.e., the press stroke and ejection of the processed article). The visualizations are presented in Figure 2.37.

The housings visible on the left-hand side of the visualizations were boxes into which articles processed by the press were ejected. The press itself was situated behind the doors of the housing visible in the central and right-hand side of the visualization images. Therefore, the beamforming method worked well in the case of impulse noise.

The last example presented in this chapter concerns examination of noise emitted by a Diesel combustion engine carried out with the use of the beamforming method and the scanning method. The engine was an element of an industrial workstation and was operated at a constant speed of 900 rpm. An electric motor (mounted below the control desk) was used to provide a load to the combustion engine and stabilize its speed. The workstation setup is shown in Figure 2.38.

In view of the relatively small overall dimensions of the examined object, there was no need to divide the measurement area into several sub-areas. The scanning procedure was realized in the form of a sequence of measurements and lasted for 30 min. Figure 2.39 shows trajectories along which the probe was led in the course of the test.

Application of the beamforming method was based on measurements lasting for 8 s and taken with the microphone array positioned parallel to the engine. The first step consisted in carrying out frequency analyses of sound pressure values acquired with the use of S&P 3D system and the microphone array in the range from 3.15 Hz to 8 kHz. Results of both analyses are presented on the same graph (Figure 2.40) in order to point out the components dominating the noise spectrum.

Subsequent plots (Figures 2.41 and 2.42) show example visualizations of the sound intensity vector and the sound pressure distribution patterns for the noise emitted by fragments of the workstation with compression-ignition engine.

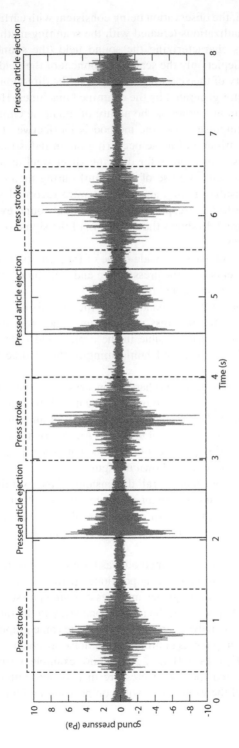

FIGURE 2.36 The waveform of signal from the microphone array placed near the press with marked fragments corresponding to the press stroke and pressed article ejection.

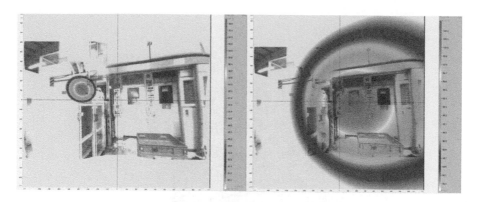

FIGURE 2.37 Visualizations of the sound pressure distribution for the events consisting in ejection of a pressed article (left) and the press stroke (right).

FIGURE 2.38 A view of a fragment of industrial workstation with compression-ignition engine.

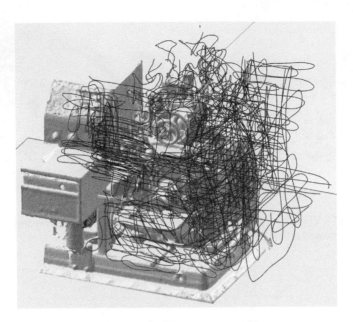

FIGURE 2.39   The probe trajectory traveled in the course of examination of an industrial workstation with compression-ignition engine.

The visualizations obtained for the above studies correspond with each other as far as locations of main noise sources are concerned. In case of the noise spectrum component with the frequency of 100 Hz, the noise source is the air filter situated above the combustion engine. For the 1700-Hz component, the mobile setup points at the right-hand side of the workstation with the diesel engine is identified as the main noise source. The noise-generating element is situated behind a protective grille. Application of the beamforming method also indicates that this element is the main source of noise; however, the obtained sound field image includes some features corresponding to reflections which can lead to false interpretation of the visualization.

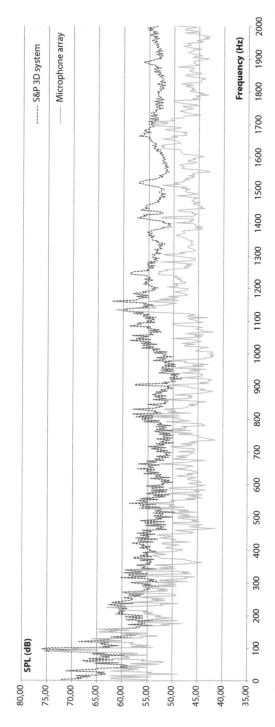

FIGURE 2.40 Spectra of signals recorded in the course of examination of a fragment of an industrial workstation with compression-ignition engine with the use of two measuring systems.

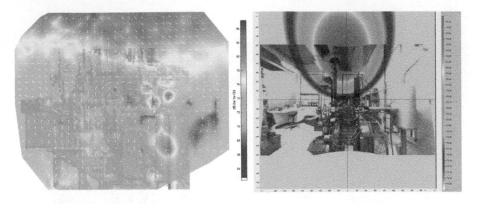

FIGURE 2.41   The sound intensity vector distribution pattern (obtained with the use of the scanning method) and the sound pressure level distribution pattern (obtained with the use of the beamforming method) for the frequency of 100 Hz.

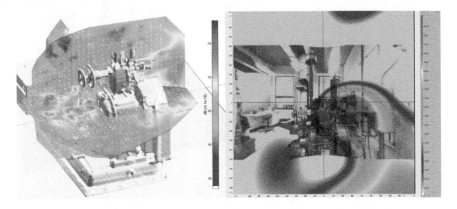

FIGURE 2.42   The sound intensity vector distribution pattern (obtained with the use of the scanning method) and the sound pressure level distribution pattern (obtained with the use of the beamforming method) for the frequency of 1700 Hz.

2.4   SUMMARY

One method for reduction of exposure to noise at workstations is to eliminate it directly at the sources which, in most cases, are machines installed in the area surrounding the workstation. The first stage of any project aimed at silencing a machine consists in identification of noise emission sources. Visualizations of parameters characterizing the sound field in the space around the machine are appropriate for that purpose. The above-presented methods of visualization of sound field parameters are used primarily to identify and assess noise emission sources. Examples of the application of individual measuring methods in modern studies on noise emission were shown. Example measuring systems were presented enabling acquisition of data for visualization of sound field parameters, including the acoustic chamber microphone system developed and installed in the Central Institute for Labour

Protection—National Research Institute. The main portion of the chapter was devoted to examples of visualization of parameters characterizing noise emitted by noise sources obtained as a result of studies carried out in laboratory conditions and as a portion of measurements taken in industrial conditions. The main advantage of the scanning method was shown to be that the visualized sound field parameters are quantities physically measured at given measurement points, rather than results of mathematical calculations. The obtained visualizations (in the form of sound intensity vector distribution patterns) are relatively easy to interpret and allow detection of possible mistakes made in the course of measuring sessions. From the material presented above, the scanning method is characterized by the longest measurement realization time and the scope of its application is seen to be best limited to stationary noise conditions. The beamforming method assumes certain simplifications (such as, for instance, the assumption that the sound waves reaching the microphone array are plane waves). The method is also sensitive to sound reflections and thus may lead to some false readings in the obtained visualizations (as was seen in the case of noise emission from a combustion engine where reflections manifested on sound pressure level distribution patterns). The beamforming method should be used in free field conditions. Correct approach to data analysis and certain level of experience in interpretation of the obtained data are indispensable to successful use of the beamforming method. The near field acoustic holography is also a technique based on transformation of measurements taken with the use of microphone array by means of mathematical operations; however, the results are characterized by a high level of consistency with those obtained from the scanning method (parameters measured directly in the sound field). Another advantage of the beamforming method and the acoustic holography technique is that they can both be used to measure and analyze impulse sound fields.

REFERENCES

Allozola, P., and B. Vonrhein. 2019. Noise source localization in industrial facilities. *Proceedings of the INTER-NOISE and NOISE-CON Congress and Conference*, June 16–19, 2019, Madrid, Spain, pp. 6997–7995.

Athanossios, M. 2015. *Beamforming Sensor Signal Processing for Defence Applications.* London: Imperial College Press.

Attendau, J.-M., and A. Ross. 2016. Time-domain near-field acoustical holography as a means to prevent exposition to harmful industrial impact noises. *Proceedings of the XXIV ICTAM*, August 21–26, 2016, Montreal, Canada.

Bai, M. R., and J. Lee, 1998. Industrial noise source identification by using an scoustic beamforming system. *J Vib Acoust-Trans ASME* 120(2):426–433. DOI:10.1115/1.2893847.

Ballesteros, J. A., E. Sarradj, M. D. Fernandez, and T. Geyer. 2015. Noise source identification with beamforming in the pass-by of a car. *Appl Acoust* 93:106–119. DOI:10.1016/j.apacoust.2015.01.019.

Barsikow, B. 1996. Experiences with various configurations of microphone arrays used to locate sound sources on railway trains operated by the DB AG. *J Sound Vibr* 193:283–293. DOI:10.1006/jsvi.1996.0269.

Barsikow, B., W. King, and E. Pfizenmaier. 1987. Wheel rail noise generated by a high-speed train investigated with a line array of microphones. *J Sound Vibr* 118(1):99–122. DOI:10.1016/0022-460X(87)90257-4.

Bolton, S. J., H. Kook, G. B. Mobes, and P. Davies. 2000. An efficient procedure for visualizing the sound field radiated by vehicles during standardized passby tests. *J Sound Vibr* 233(1):137–156. DOI:10.1006/jsvi.1999.2794.

Bonanomi, A. 2017. Acoustic analysis of electric motors in noisy industrial environment. tech. rep. Power Transmission World. www.powertransmissionworld.com/acoustic-analysis-of-electric-motors-in-noisy-industrial-environment. (accessed January 28, 2020).

Brooks, T. F., and W. M. Humphreys. 2004. A deconvolution approach for the mapping of acoustic sources (DAMAS) determined from phased microphone arrays. *Proceedings of the 10th AIAA/CEAS Aeroacoustics Conference*, May 10–12, 2004, Manchester, UK: AIAA Paper 2004-2954.

Brühl, S., and K. P. Schmitz. 1993. Noise source localization on high-speed trains using different array types. *Proceedings of the Internoise 93*, August 24–26, 1993, Leuven, Belgium, pp. 1311–1314.

Buck, S., S. Oerlemans, and S. Palo. 2016. Experimental characterization of turbulent inflow noise on a full-scale wind turbine. *J Sound Vibr* 385:219–238. DOI:10.1016/j.jsv.2016.09.010.

Chelliah, K., and G. G. Raman. 2014. Advanced aeroacoustic testing techniques using various methods of acoustic holography. *Proceedings of the 20th AIAA/CEAS Aeroacoustics Conference*, June 16–20, 2014. Atlanta, GA: American Institute of Aeronautics and Astronautics.

Chelliah, K., G. G. Raman, R. T. Muehleisen, H. Patel, and E. Tatara. 2015. Building leakage detection and quantification using statistically optimized nearfield acoustic holography technique. *J Acoust Soc Am* 137: 2325–2325. DOI:10.1121/1.4920491.

Cho, Y. T. 2015. Noise source identification of reciprocating air compressor in a normal operating condition using spatially low-resolution sound measurement. *Indian J Sci Technol* vol. 8. https://pdfs.semanticscholar.org/2036/98107d507d18228685c6fcf0e02d d720a2dc.pdf. (accessed January 28, 2020).

Christensen, J. J. 2004. Technical review: Beamforming. Tech. Rep., Brüel & Kjær. www.bksv.com/-/media/literature/Technical-Review/bv0056.ashx. (accessed January 28, 2020).

Chyariottia, P., M. Martarellia, E. P. Tomasinia, and R. Beniwalb. 2010. Noise source localization on washing machines by conformal array technique and near field acoustic holography. *Proceedings of the IMAC-XXVIII*. February 1–4, 2010, Jacksonville, FL: Society for Experimental Mechanics Inc.

de la Croix D. V. U., D. Webster, B. Gariner, and F. Molliex. 1997. Malice the efficient acoustic imaging system for precise noise source localization. *Proceedings of 5th International Congress on Sound and Vibration*, December 15–18, 1997, South Australia.

Derouiche, A., N. Hamazoui, and T. Boukharouba. 2012. Reconstruction of sound sources of gear transmission mechanism by planar near field acoustical holography. In *Condition Monitoring of Machinery in Non-Stationary Operations*, Berlin: Springer, pp. 247–255.

Dougherty, R., and B. Walker. 2009. Virtual rotating microphone imaging of broadband fan noise. *Proceedings of 15th AIAA/CEAS Aeroacoustics Conference (30th AIAA Aeroacoustics Conference)*, May 11–13, 2009, Miami, FL: American Institute of Aeronautics and Astronautics.

Hald, J. 2001. Time domain acoustical holography and its applications. *Sound Vib* 35(2):16–35.

Hayek, S. I. 2008. Nearfield acoustical holography. In *Handbook of Signal Processing in Acoustics*, eds. D. Havelock, S. Kuwano, M. Vorländer, pp. 1129–1139. New York: Springer. DOI:10.1007/978-0-387-30441-0_59

He, B., and X. S. Jin. 2014. Investigation into external noise of a high-speed train at different speeds. *J Zhejiang Univ Sci A* 15:1019–1033.

Hou, J., W. Jiang, and W. Lu. 2011. Application of a near-field acoustic holography-based diagnosis technique in gearbox fault diagnosis. *J Vib Control* 19(1):3–13. DOI:10.1177/1077546311428634.

Howell, G., A. Bradley, M. McCormick, and J. Brown. 1986. De-dopplerization and acoustic imaging of aircraft flyover noise measurements. *J Sound Vibr* 105(1):151–167. DOI:10.1016/0022-460X(86)90227-0.

Johnson, D. H., and E. D. Dudgeon. 1993. *Array Signal Processing: Concepts and Techniques.* Englewood Cliffs: Prentice-Hall.

Kim, Y.-H., and J.-W. Choi. 2013. *Sound Visualization and Manipulation.* Singapore: Wiley.

King, W., and D. Bechert. 1979. On the sources of wayside noise generated by high-speed trains. *J Sound Vibr* 66(3):311–332. DOI:10.1016/0022-460X(79)90848-4.

Kitagawa, T., and D. Thompson. 2010. The horizontal directivity of noise radiated by a rail and implications for the use of microphone arrays. *J Sound Vibr* 329(2):202–220. DOI:10.1016/j.jsv.2009.09.002.

Kook, H., P. Davies, and S. J. Bolton. 1999. The design and evaluation of microphone arrays for the visualization of noise sources on moving vehicles. *SAE Technical Paper* 1999-01-1742. DOI:10.4271/1999-01-1742.

Kreider, W., P. V. Yuldashev, O. A. Sapozhnikov et al. 2013. Characterization of a multi-element clinical HIFU system using acoustic holography and nonlinear modeling. *IEEE Trans Ultrason Ferroelectr Freq Control* 60(8):1683–1698. DOI:10.1109/TUFFC.2013.2750.

Lafon, B., and J. Antoni. 2008. Cyclic sound intensity and source separation from NAH measurements on a diesel engine. *J Acoust Soc Am* 123:3387–3387. DOI:10.1121/1.2934044.

Lee, M., and S. J. Bolton. 2007. Source characterization of a subsonic jet by using near-field acoustical holography. *J Acoust Soc Am* 121(2):967–977.

Lee, M., S. J. Bolton, and L. Mongeau. 2003. Application of cylindrical near-field acoustical holography to the visualization of aeroacoustic sources. *J Acoust Soc Am* 114:842–858. DOI:10.1121/1.1587735.

Lu, M.-H., and M. U. Jen. 2010. Source identification and reduction of engine noise. *Noise Control Eng J* 58(3):251–258. DOI:10.3397/1.3427147.

Lu, W., W. Jiang, G. Yuan, and L. Yan. 2013. A gearbox fault diagnosis scheme based on near-field acoustic holography and spatial distribution features of sound field. *J Sound Vibr* 332(10):2593–2610. DOI:10.1016/j.jsv.2012.12.018.

Luo, Z.-W., D. Fernandez Comesaña, C.-J. Zheng, and C.-X. Bi. 2019. Near-field acoustic holography with three-dimensional scanning measurements. *J Sound Vibr* 439:43–55. DOI:10.1016/j.jsv.2018.09.049.

Mann III, J. A., and J. C. Pascal. 1992. Locating noise sources on an industrial air compressor using broadband acoustical holography from intensity measurements (BAHIM). *Noise Control Eng J* 39(1):1–3.

Maynard, J. D., E. G. Williams, and Y. Lee. 1985. Nearfield acoustic holography: 1 theory of generalized holography and the development of NAH. *J Acoust Soc Am* 78(4):1395–1413. DOI:10.1121/1.392911.

Merino-Martinez, R., M. Snellen, and D. G. Simons. 2016a. Functional beamforming applied to full scale landing aircraft. *Proceedings of the 6th Berlin Beamforming Conference,* 29 February–1 March 2016, Berlin, Germany.

Merino-Martinez, R., M. Snellen, and D. G. Simons. 2016b. Functional beamforming applied to imaging of flyover noise on landing aircraft. *J Aircr* 53:1830–1843. DOI:10.2514/1.C033691.

Michel, U., B. Barsikow, P. Bohning, and M. Hellmig. 2004. Localisation of moving sound sources with phased microphone arrays. *Proceedings of INTER-NOISE and NOISE-CON Congress and Conference,* August 22–25, 2004, Prague, Czech Republic, pp. 4069–4075.

Morzyński, L., and G. Szczepański. 2018. Double panel structure for active control of noise transmission. *Arch Acoust* 43(4):689–696. DOI:10.24425/aoa.2018.125162.

Oerlemans, S., P. Sijtsma, and M. B. López. 2007. Location and quantification of noise sources on a wind turbine. *J Sound Vibr* 299(4–5):869–883. DOI:10.1016/j.jsv.2006.07.032.

Orman, M., and C. T. Pinto. 2013. Acoustic analysis of electric motors in noisy industrial environment. *Proceedings of 12th IMEKO TC10 Workshop on Technical Diagnostics New Perspectives in Measurements, Tools and Techniques for Industrial Applications*, June 6–7, 2013, Florence, Italy.

Pannert, W., and C. Maier. 2014. Rotating beamforming – Motion-compensation in the frequency domain and application of high-resolution beamforming algorithms. *J Sound Vibr* 333:1899–1912. DOI:10.1016/j.jsv.2013.11.031.

Patel, H. J., K. Chelliah, G. Raman, R. T. Muehleisen, and E. Tatara. 2015. Detecting building leakages using nearfield acoustic holography technique: A numerical simulation. *J Acoust Soc Am* 137:2233–2233. DOI:10.1121/1.4920147.

Peterson, J. S., D. Yantek, and A. K. Smith. 2012. Acoustic testing facilities at the office of mine safety and health research. *Noise Control Eng J* 60(1):85–96. DOI:10.3397/1.3678441.

Pleban, D. 2012. *Jakość akustyczna maszyn* [Acoustic quality of machines]. Warszawa: Centralny Instytut Ochrony Pracy – Państwowy Instytut Badawczy.

Sapozhnikov, O. A., S. A. Tsysar, V. A. Khokhlova, and W. Kreider. 2015. Acoustic holography as a metrological tool for characterizing medical ultrasound sources and fields. *J Acoust Soc Am* 138(3):1515–1532. DOI:10.1121/1.4928396.

Schwartz, O., S. Gannot, and E. A. P. Habets. 2015. Nested generalized sidelobe canceller for joint dereverberation and noise reduction. *2015 IEEE International Conference on Acoustics, Speech and Signal Processing (ICASSP)*, April 19–24, 2015, Brisbane, Australia.

Serraris, J. 2016. Propagation of sound around different noise barrier models using a three-dimensional scan based sound visualization technique. *Noise Control Eng J* 64(2):134–141. DOI:10.3397/1/376366.

Sijtsama, P. 2010. Phased array beamforming applied to wind tunnel and fly-over tests. *SAE Technical Paper Series, SAE International*, October 17–19, 2010, Florianópolis, Brasil. https://reports.nlr.nl/xmlui/bitstream/handle/10921/192/TP-2010-549.pdf. (accessed January 28, 2020).

Simons, D. G., M. Snellen, B. van Midden, M. Arntzen, and D. H. T. Bergmans. 2015. Assessment of noise level variations of aircraft flyovers using acoustic arrays. *J Aircr* 52:1625–1633. DOI:10.2514/1.C033020.

Sklanka, B., J. Tuss, R. Buehrle, J. Klos, E. Williams, and N. Valdivia. 2006. Acoustic source localization in aircraft interiors using microphone array technologies. *Proceedings of the 12th AIAA/CEAS Aeroacoustics Conference (27th AIAA Aeroacoustics Conference)*, May 8–10, 2006, Cambridge, MA: American Institute of Aeronautics and Astronautics.

Takano, Y., K. Tereda, E. Aizawa, A. Iida, and H. Fujitta. 1992. Development of a 2-dimensional microphone array measurement system for noise sources of fast moving vehicles. *Proceedings of Internoise 92*, July 20–22, 1992. Toronto, ON, pp. 1175–1178.

Thompson, D. J., E. L. Iglesias, X. Liu, J. Zhu, and Z. Hu. 2015. Recent developments in the prediction and control of aerodynamic noise from high-speed trains. *Int J Rail Transp* 3(3):119–150. DOI:10.1080/23248378.2015.1052996.

Wall, A. T., K. L. Gee, T. B. Neilsen, D. W. Krueger, and M. M. James. 2014. Cylindrical acoustical holography applied to full-scale jet noise. *J Acoust Soc Am* 136:1120–1128. DOI:10.1121/1.4892755.

Wall, A. T., K. L. Gee, T. B. Neilsen, R. L. McKinlez, and M. M. James. 2016. Military jet noise source imaging using multisource statistically optimized near-field acoustical holography. *J Acoust Soc Am* 139:1938–1950. DOI:10.1121/1.4945719.

Weyna, S. 1995. The application of sound intensity technique in research on noise abatement in ships. *Appl Acoust* 44(4):341–351. DOI:10.1016/0003-682X(94)00031-P.

Weyna, S. 2003. Identification of reflection, diffraction and scattering effects in real acoustic flow fields. *Arch Acoust* 28(3):191–203.

Weyna, S. 2005. *Rozpływ energii akustycznych źródeł rzeczywistych* [Energy flow patterns of real sound sources]. Warszawa: Wydawnictwa Naukowo-Techniczne.

Williams, E. G. 1999. *Fourier Acoustics: Sound Radiation and Nearfield Acoustical Holography.* San Diego: Academic Press.

Williams, E. G., B. H. Houston, P. C. Herdic, S. T. Raveendra, and B. Gardner. 2000. Interior near-field acoustical holography in flight. *J Acoust Soc Am* 108(4):1451–1463. DOI:10.1121/1.1289922.

Yang, B. 2013. Research status on aero-acoustic noise from wind turbine blades. *IOP Conf Ser: Mater Sci Eng* 52(1). DOI:10.1088/1757-899X/52/1/012009.

Yang, D., Z. Wang, B. Li, Y. Luo, and X. Lian. 2011. Quantitative measurement of pass-by noise radiated by vehicles running at high speeds. *J Sound Vibr* 330(7):1352–1364. DOI:10.1016/j.jsv.2010.10.001.

Yantek, D. S., J. S. Peterson, and A. K. Smith. 2007. Application of a microphone phased array to identify noise sources on a roof bolting machine. *Proceedings of the Noise-Con 2007*, October 22–24, 2007, Reno, Nevada.

3 The Surround Sound in Aural Perception Tests

Grzegorz Szczepański

CONTENTS

3.1 INTRODUCTION

The human ear is the second most important sensory organ, next only to the eye, by means of which workers acquire information about the work environment. A human in the work environment is subject to a plurality of different sound stimuli, a portion of which constitute the source of information necessary to carry out the work. The stimuli include, among other things, verbal messages exchanged with co-workers and auditory signals generated by machines (e.g., alarm buzzers), and are a complement to visually perceived information about conditions prevailing in the work environment and potential hazards, especially in regions outside the field of vision. The remaining portion of the sound environment is noise which not only makes it difficult to work but in many cases has a harmful effect on human health. In today's work environment, guaranteeing that workers will be able to perceive vital sound signals in diversified conditions, must be considered equally important to reduction of exposure to noise hazards. Such signals, essential from the point of view of both safety and correctness of the performed work, are verbal messages issued by operators and co-workers, communications, and auditory warning systems installed in the workplace.

In the case of direct messages and general communications, the most important parameter decisive for correct reception of the content is the speech intelligibility which may be affected by parameters of the listening room—in particular the reverberation time (depending on the room size and sound absorption properties), the background sound pressure level, and the parameters of the speech signal itself. Where acoustic warning systems are used, it is of great importance that the quality of system components (such as microphones and loudspeakers) is adequate. To date, numerous test methods have been developed and used to assess intelligibility of speech in the work environment. However, studies on recognizability of auditory danger signals pertain to a somewhat different range of issues. Provisions concerning properties which should characterize any danger signal in Europe is currently contained in the standard EN ISO 7731:2008. This standard introduces the notion of the auditory danger signal as a "signal marking the onset and, if necessary, the duration and the end of a dangerous situation at a workplace." Every person present in the danger signal reception area should be able to recognize it, therefore the signal must be clearly audible. Usually, such signals are amplitude- or frequency-modulated sounds of a basically tonal nature.

Studies on perception of sound signals (including danger signals) should include the correct identification of actual meaning of the signal and correct perception of the direction from which it emanates. These studies are usually carried out in laboratory conditions. The test signals and reproduction of the sound field (the listening space prevailing at actual workstations) are simulated with the use of surround sound reproducing systems. The process of creation of such simulated virtual acoustic environments is commonly called *auralization*. A plurality of auralization techniques is now available. In each, achievement of the surround sound effect is a multi-stage process involving recording a sound scene or the sound emitted by a specific device, then processing the recorded signals, and finally playing them back on a laboratory setup. Usually, the surround sound–reproducing technique is in direct relation to the sound recording methods. Research on (and development of) auralization techniques are of fundamental importance in studies concerning the assessment of hearing directivity in workers (their ability to identify which direction a sound is coming from) within the work environment and for predictions concerning acoustic properties of buildings in early stages of developing the construction design.

This chapter characterizes briefly various methods used to create virtual acoustic environments from the point of view of application in research projects concerning reception of acoustic signals. Further, the structure of a laboratory auralization setup is presented in which the ambisonic technique was used to create virtual acoustic environments. The chapter ends with a summary of results of example tests concerning sound signal reception, especially in visually impaired persons.

3.2   VIRTUAL ACOUSTIC ENVIRONMENT REALIZATION METHODS

Studies on perception of sound signals are typically carried out in laboratory setups designed and constructed specifically for research purposes. Perception test results are therefore closely connected with the degree to which an artificially created virtual acoustic environment is capable of reproducing a sound field actually existing in specific real-life conditions. To create a virtual acoustic environment, it is necessary to record sound signals in an actual environment and then reconstruct them with the use of a playback system. Other matters that need to be decided include processing algorithm selection and realization of the actual listening session (surround sound reproducing systems or headphones), the listening room, and parameters of individual components used to create a virtual acoustic environment.

The oldest known methods of obtaining the sensation of surround sound are stereophonic techniques [Williams 1987]. In order to play back a sound with the use of stereophony, it is necessary to make use of one of many well-known sound recording methods [Zhang et al. 2017] such as AB, XY, or ORTF and their combinations. The methods are useful in the case of sound reproduction in surround stereophonic systems and are subject to different limitations (such as absence of time lag between channels in the XY method). The range of surround effects that can be obtained in the course of playing back these recordings is very limited. An extension of the classical stereophony over three spatial dimensions is the so-called vector-base amplitude panning (VBAP) function developed in the 1990s [Pulkki 1997]. This method enables the generation of three-dimensional sound field effects from loudspeakers arranged on the projection of a triangle.

Another technique enabling the creation of a virtual acoustic environment in surround systems is the ambisonic technique. The invention of the ambisonic technique is attributed to Michael Gerzon who published his key works on the subject at the turn of the 1970s and 1980s [Gerzon 1973, 1974]. The technique enables one to record, process, and reconstruct sound fields in a fully three-dimensional space. Recording is carried out with the use of an ambisonic microphone comprised of a definite number of component microphones. A first-order ambisonic microphone consists of four microphones arranged on the faces of a regular tetrahedron; an example of such microphone is presented in Figure 3.1.

The ambisonic technique is based on processing signals coming from a number of microphones into a virtual single (central) point source situated between the microphones, with phase shifts due to separation of component microphones taken into account. In the case of first-order ambisonics, this consists of four signals—of which one carries omnidirectional information, while the other three components

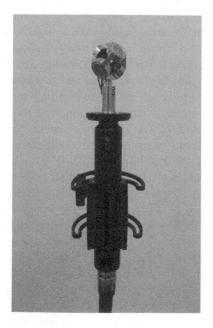

FIGURE 3.1 A first-order ambisonic microphone.

are proportional to pressure gradients connected with distribution of signals incoming from three orthogonal directions. The description is expressed in terms of spherical harmonics where a larger number of microphones makes it possible to use higher-order spherical harmonic functions. This translates into improvement of the obtained sound source location resolution [Bertet et al. 2013]. Currently used sound transducers include both first-order ambisonic microphones [Carlsson 2004; Clapp et al. 2014] and microphones of higher orders [Satongar et al. 2013; Vennerod 2014].

Recordings registered using an ambisonic microphone can be played back on surround sound systems of various types and with different numbers of loudspeakers, therefore the recorded signal must be decoded in connection with data-carrying information on a particular arrangement and number of loudspeakers. Currently, the market offers a wide choice of dedicated programs and software VST (Virtual Studio Technology) plug-ins facilitating the process of developing ambisonic decoders for loudspeaker arrangements defined by the user. Results of some studies [Heller et al. 2008; Moore and Wakefield 2008] show that the type of ambisonic decoder has an effect on the achieved reproduction quality of the virtual acoustic environment. Another element necessary for a virtual acoustic environment is a surround sound reproducing system. Directional resolution of reproduced signals also depends on the number and arrangement of loudspeakers. The number of loudspeakers must not be too small, as that results in poor resolution; however, an excessively large number of loudspeakers generates irregularities in the signal spectrum. In cases where these spectrum irregularities are encountered, graphical spectrum correctors (equalizers) are typically used. These apply filtration to spectrum components detected as inconsistent with respect to recordings taken in actual conditions [Yao 2018]. The number of sound sources (together with the number of their possible locations) lead to an infinite number of theoretically possible configurations of virtual acoustic setups, from standardized arrangements [Recommendation 2017] to dedicated application-specific solutions [Epain et al. 2014]. Sound reproduction is realized for a defined listening area situated near the setup center point, but leaning one's head out of that region results in deterioration of hearing impressions and problems with determination of the sound source location [Stitt et al. 2014]. Apart from the number and arrangement of loudspeakers, another matter of importance might be their frequency characteristics, as well as other parameters. The last element having an effect on realization of a virtual acoustic environment are the parameters of the listening room [Sawicki and Mickiewicz 2006; Harley 2010; Pelzer et al. 2014]. Even a correctly recorded signal, in combination with an excellent surround sound system, will not enable faithful reproduction of the actual acoustic conditions if the listening room is not prepared properly. The listening room must be a space with appropriate acoustic absorption and must be effectively isolated from ambient sounds (especially in case of reproduction of virtual acoustic environments with low sound pressure levels). These properties of the listening room depend on many factors, including the arrangement and type of structural materials, as well as the shape and size of the room itself. If a listening room lacks any acoustic adaptation and is small in size, it is possible to make use of reflections as additional sound sources [Betlehem et al. 2010].

In discussing the creation of virtual acoustic environments, it is also necessary to mention the sound wave synthesis technique [Rabenstein et al. 2014] and the binaural

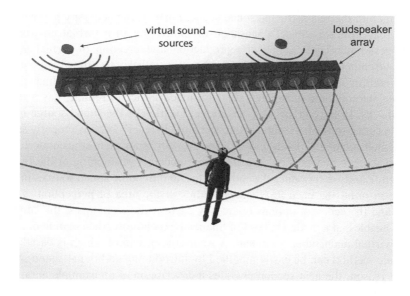

FIGURE 3.2   The principle of the sound wave synthesis.

technique [Hammershoi and Møller 2002]. The sound wave synthesis technique uses large numbers (tens or hundreds) of individually controlled loudspeakers positioned close to each other to create an "artificial" sound wavefront. The sound wavefront created in this manner should correspond to the actual sound wave produced by sound sources placed at specific points in space. The principle of this method is illustrated in Figure 3.2. The sound field synthesis technique enables the creation of virtual acoustic environments in spaces with larger volumes. By contrast with other techniques in which the sound field is controlled at a specific point in space and aural sensations depend on the listener's position relative to that point, the sound wave synthesis technique makes it possible to obtain a homogeneous listening area. However, application of the method requires employment of elaborate sound reproducing systems and digital signal processing.

In the binaural technique, acoustic signals are recorded using systems of "dummy head" type or binaural microphones. Such recording methods use the fact that the listener's body parts (head, torso, and auricles) form a kind of an acoustic circuit called the head-related transfer function (HRTF) filter. The filtration modifies the frequency structure of sound reaching the listener's ears. Sound reproduction with the use of the binaural technique occurs by supplying two independent signals to each ear via headphones [Assenmacher et al. 2004]. A problem connected with this technique is the absence of a link between the reproduced signals and moves of the listener's head. This issue is the subject of several specific studies [Deldjoo and Atani 2016; Echevarria et al. 2017].

Each of these techniques has its own flaws and merits. In the binaural technique, the necessity to use headphones is an obstacle in studies where the subjects wear hearing protectors. It is also difficult to modify acoustic images once they are recorded, perhaps to introduce additional warning signals coming from various directions into the reconstructed sound scene. Further, the necessity to wear head-phones may affect a subject's sensation of immersion in a virtual environment. The largest potential in scope of reproduction and creation of virtual acoustic work environment is offered by the sound wave synthesis technique, although the method is difficult to implement in practice and requires considerable financial out-lays. The stereophonic technique, although much easier to realize, has a number of drawbacks. In the stereophonic technique, location of the sound source is suffi-ciently accurate in the horizontal plane, when the listener's head is properly situated and oriented relative to loudspeakers. However, in other planes, accurate location of sound sources becomes much worse. Any incorrect positioning of the listener's head significantly deteriorates the ability to locate individual sound sources in space. Bearing in mind the need to continue studies in the area of perception of sound signals and the scope of options offered by each of the techniques, the ambisonic method enables, at a moderate level of financial expenditure, the creation of a good-quality virtual acoustic environment. A setup, the design of which is based on the ambisonic method can be easily modified by introducing additional sound sources. For that reason, the next section provides a description of an example auralization setup in which the ambisonic technique is employed.

3.3   AN AMBISONIC TECHNIQUE-BASED AURALIZATION SETUP

A setup for surround auralization with the use of the ambisonic technique can be realized in many different ways. The approach to creation of virtual acoustic work environments presented in this chapter (together with examples of equipment and configurations of the setup) was realized at the Central Institute for Labour Protection—National Research Institute in Warsaw, Poland. Any setup of that kind should enable the reproduction of sound fields of selected actual locations as pre-cisely as possible (with special emphasis put on workstations) and the creation of acoustic scenarios meeting the needs of specific auditory tests. Generally, the factors producing an actual effect on quality of the final result can be divided into three groups. These are factors connected with:

- the system recording an actual environment,
- the listening room, and
- the surround sound playback system.

The basic component of any virtual acoustic environment is a sound scene which constitutes a sound recording of a given environment (neighborhood) recorded with the use of a recording system comprised of a suitable microphone and recorder. Recall that the resolution of the sound field reconstruction process in the ambisonic technique is affected by the order of microphones used for signal recording. In situa-tions where sound field reconstruction is undertaken solely to create a sonic image of a scene and immerse a subject in a specific reality and sound stimuli are reproduced separately from additional monophonic channels, the resolution is not the matter of key importance for achieving the test objective. In such cases, the role of the ambisonic technique is to create an ambient sound for a given test. For that reason, recordings were carried out with the use of a first-order ambisonic microphone AMBEO VR Mic (Sennheiser), cooperating with Tascam DR-680 MkII multi-channel digital recorder. In order to acquire necessary information about both the environment in which the sound recording was made and acoustic objects and events reproduced in the recording, it is necessary to complement the sound recording with properly synchronized video documentation. A good solution would be to extend the sound recording system by coupling it with a camera recording all-around images (a 360° video camera). Such cameras are commonly available and are characterized by small dimensions and light weight which is an important feature due to the necessity of placing them on stands close to ambisonic microphones (the camera should not distort the sound field in the course of recording sessions). A view of such a set of sound- and image-recording devices is shown in Figure 3.3.

In view of the fact that a 3D camera records the image with the use of two lenses, the output videos must be subjected to processing. There are multiple options of processing the source files of camera recordings including, among other things, a twin screen (stretching the front and the rear component of the video over the whole frame), a panoramic view, and a view in the 360° mode. For the purpose of identification of sources and actual directions from which the sound is coming, the videos can be converted into 360° mode which enables playback of the videos using virtual reality goggles and dedicated video players. A view of the converted video enables correct interpretation of different situations occurring at a given moment of time. Example video frames are shown in Figure 3.4.

A precondition of realization of a good-quality virtual acoustic environment includes proper adaptation of the listening room. The auralization setup was

FIGURE 3.3   An ambisonic microphone and a 360° video camera.

FIGURE 3.4   Different views obtained from a single 360° video frame.

placed in an acoustic chamber offering conditions close to those of the free field with a sound-reflecting floor. The chamber floor was covered with sheet flooring. The ceiling of the chamber was a self-supporting structure made of many layers of multi-functional plate (MFP) and gypsum-cardboard panels. The room was also subjected to acoustic treatment consisting in that its walls and the ceiling were covered with sound-absorbing structures based on two types of sound-absorbing foam elements in the form of wedges. Inner dimensions of the chamber after adaptation were 13.5 m × 4.4 m × 5.0 m (length × width × height). Net cubature of the chamber after adaptation was 297 m³. A view of the listening room is shown in Figure 3.5.

FIGURE 3.5   The listening room with the auralization setup.

Reconstruction of surround sound at a laboratory listening setup based on recordings obtained with the use of an ambisonic microphone requires that the registered sound signals are subject to preprocessing. The first step consists in conversion of signals recorded by individual microphone capsules making up a tetrahedral microphone (the so-called A-format) to a format in which individual signals correspond to recordings obtained from a microphone with omnidirectional directivity pattern situated at a point and three orthogonally oriented microphones with the figure-eight directivity pattern (the so-called B-format). The sound information coded in B-format must be then transformed (decoded) into n-channel sound, transmitted by a set of n loudspeakers, each with a defined position in space, allowing reconstruction of the recorded sound field in laboratory conditions. To decode sound recorded in the ambisonic format, it is possible to use commercial versions of software, including VST plug-ins installed in software of digital audio workstation (DAW) type. Literature of the subject offers descriptions of toolkits for signal decoding to be used for setups with irregular distribution of loudspeakers [Heller et al. 2010, 2012]. In the following example, Nuendo version 6 software (Steinberg) was used for signal processing and a Sennheiser AMBEO A-B converter plug-in was employed to convert recordings from A-format to B-format.

The process of surround sound reproduction in the laboratory setup was realized based on a multi-channel system offering 128 individually controlled output channels. Schematic diagram of the system is presented in Figure 3.6.

The system is operated with the use of Dante (Digital Audio Network Through Ethernet) standard in which digital acoustic signals are transmitted via Ethernet.

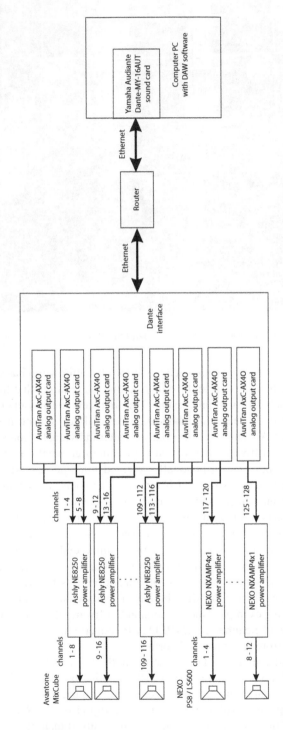

FIGURE 3.6   Schematic diagram of the multi-channel system in the acoustic test chamber.

The system uses Audiante Dante-MY-16AUT sound card installed in a PC-class computer and uses one hundred and sixteen Avantone MixCube studio monitors, six Nexo PS8 stage monitors, and six LS600 woofers.

Subjective aural sensations and the quality of sound displayed within a virtual acoustic environment may be assessed in different ways for different sound scenes, according to the results of studies carried out on a variety of reproduction systems using the ambisonic technique to playback musical material [Marentakis et al. 2014]. To construct a surround sound reproducing system, properly selected and arranged elements of output channels of the multi-channel system were used. The setup comprised 19 loudspeakers in total, including 17 studio monitors and 2 woofers. The studio monitors were distributed over the surface of a sphere with the radius of 2 m (the center of the sphere coinciding with the location of the listener's head during the testing) in the following way:

- eight studio monitors distributed evenly on a circle (45° apart) in the horizontal plane at the listener's head height,
- four studio monitors set out evenly on a circle (90° apart, with loudspeaker axes oriented at the angle of 45° relative to the front-back axis) in the floor plane, pointed at the subject at the angle of 45° relative to the floor plane,
- four studio monitors distributed evenly on a circle (90° apart, with loudspeaker axes oriented at the angle of 45° relative to the front-back axis) in the plane situated above the listener and pointed at the subject at the height for which the angle between the loudspeaker axis and the horizontal plane is 45°,
- one loudspeaker is placed exactly above the subject, and
- two woofers are placed on the floor in the front hemisphere at the angle of 45° relative to the front-back axis and at the distance of 2.5 m from the subject's head.

A schematic diagram illustrating the arrangement of loudspeakers in the surround sound reproducing system set up for creating a virtual acoustic work environment is shown in Figure 3.7.

In the ambisonic technique, an integral element of the process is decoding the signal from B-format into the input signal for loudspeakers of the surround sound reproducing system. The decoder must be fed with information on positions of loudspeakers relative to the listening spot. Without a correctly configured decoder it would be impossible to carry out experiments involving the use of a given number of loudspeakers in a given arrangement. In the arrangement optimization study, Rapture 3D Advanced Speaker Layout software was used as the main ambisonic decoder. The software enables to generate an ambisonic decoder for any arrangement of loudspeakers in space. Reproduction of surround sound is effected with the use of another module of Rapture 3D software—the Rapture 3D Player. The ambisonic decoder generated by the Advanced Speaker Layout module is automatically read out by the Rapture 3D Player. Reproduction of surround sound consists in importing a recording registered in ambisonic B-format to the Rapture 3D Player program and starting the playback function.

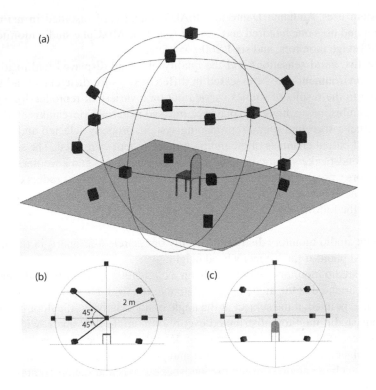

FIGURE 3.7   A schematic view of 3D arrangement of loudspeakers in the laboratory sound reproducing system—projections (a) axonometric, (b) from left, and (c) from behind.

Tests are carried out with the use of acoustic scenarios defined as sequences of acoustic events and sounds reproducing a specific actual situation in the virtual acoustic environment. Any specific acoustic scenario comprises therefore an ambient sound and additionally introduced test signals. In tests that are carried out with this laboratory setup, signals recorded in real-life conditions can be used both as test signals for experiments and as a background for test signals generated in other ways (where it is impracticable to register a required sound scenario in real life). In view of the above, the laboratory setup for realization of virtual acoustic work environments was expanded by adding a test stimuli playback system. A schematic diagram of such system is shown in Figure 3.8.

The system is based on a PC-class computer equipped with MOTU PCI-424 sound card and with Adobe Audition software installed which, via MOTU 24I/O interface, control a set of eight M-Audio BX5 D2 loudspeakers. The loudspeakers are evenly distributed along a circle at the listener's head height, just under loudspeakers of the surround sound reproducing systems. A schematic diagram of loudspeakers arrangement in the stimuli playback system is depicted in Figure 3.9.

The developed setup may be used to carry out sound perception tests [Morzynski et al. 2017; Szczepański et al. 2018] in research projects concerning, for example,

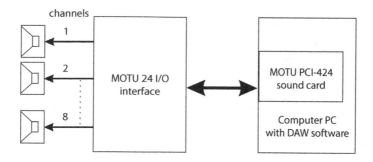

FIGURE 3.8 A schematic diagram of the test stimuli playback system.

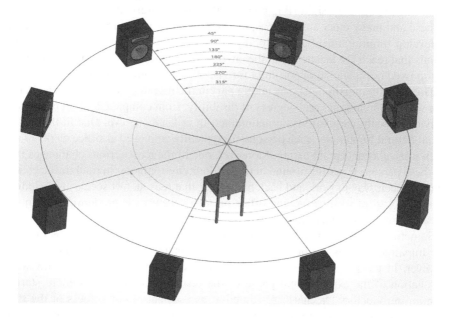

FIGURE 3.9 A schematic diagram of arrangement of loudspeakers in the laboratory stimuli playback setup with channel number and signal playback direction (angle) marked.

spatial imagination in visually impaired persons, directivity of hearing vehicle-reversing signals in workers wearing hearing protectors, and functionality of hearing protectors with attenuation adjustment [Mlynski and Kozlowski 2019].

3.4 EXAMPLE SOUND PERCEPTION TESTS CARRIED OUT WITH THE USE OF THE AURALIZATION SETUP

Subjective assessment of the quality of reproduction of an actual environment in the virtual acoustic work environment requires that the subjects asked to make such assessment are familiar with the original acoustic environment from their own experience. The present section describes a simple test procedure and results of setup testing carried out with a group of 10 males aged from 30 to 40, of whom 7 men were visually impaired with a specific level of self-reliance (individuals able to move around on their own in the public space), whereas 3 subjects were fully able-bodied persons with good vision.

Visually impaired persons constitute a specific group of workers for whom the hearing organ is the main sensory mechanism used as the source of information about the environment. Any sound existing in the work environment may be perceived by them as either a disturbing noise or the carrier of useful information decisive for correctness of their behavior and safety. Intensive research work is carried out worldwide on developing innovative devices [Dunai et al. 2015] for visually impaired persons to aid them in maneuvering in urban agglomerations. Research is also active in the area of training designed to improve spatial imagination in persons with impaired eyesight [Bălan et al. 2015] and disturbed spatial imagination. In these and many other applications, the virtual acoustic environment can be used to develop training aimed at developing correct directional assessment of specific sounds that reach the listener from the environment. Thus, this research contributes to improved occupational opportunity for visually impaired individuals.

The visually impaired persons taking part in the research project were characterized with different types and degrees of disability, from completely blind persons to individuals with marginal vision maintained in at least one eye. Due to their individual natural limitations, such persons are typically employed as office clerks or are given jobs of a conceptual nature. Moreover, one of the most serious challenges facing these persons is getting around in the outdoor environment (in road traffic conditions) on their way to work and back home. For that reason, test scenarios including sounds typical of road traffic and office premises were used predominantly in the research project presented in this section.

Sometimes, tests on perception of sound signals realized with groups of visually impaired persons are preceded by a questionnaire survey [Hojan et al. 2012]; however, in the present case it was decided to conduct such a survey only after completion of the experimental phase of the test. The research procedure started with an introductory "Scenario A" aimed at assessment of correctness of the subject's orientation relative to the sound scene. The recording used in that scenario included sounds of traffic with very low intensity taking place in a local asphalt road (Figure 3.10). The sound scene duration was three minutes. The sound sources identifiable in that scene included, among other things, motor cars passing by, an agricultural machine, and an airplane flying overhead. Figure 3.11 shows the waveform of the sound signal in which fragments corresponding to individual sound sources are marked. The signal used in Scenario A was recorded facing the road, rather close (about 1 m) to the roadway. The motor cars passed by occasionally, whereas the sound of an airplane flying by was recorded when no road traffic occurred. Therefore the sound scene turned out to be a good introductory example to assess correctness of impressions in subjects and their ability to perceive individual components of the sound environment.

The task given the subjects was to listen intently to the presented sound scene and memorize as many details concerning the reproduced sounds as possible—in particular, what were the sources of the sounds and from which directions they came.

FIGURE 3.10 A view of location site on which recordings used in Scenario A were registered.

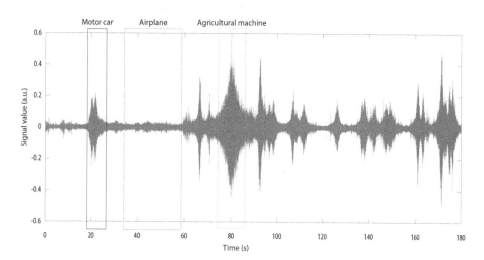

FIGURE 3.11 The waveform of the sound signal used in Scenario A with marked fragments corresponding to selected sound sources.

Detailed follow-up questions concerned the orientation of the street with respect to their position and the direction in which the airplane was flying. Table 3.1 summarizes answers given by the subjects.

Results of the task based on Scenario A indicate that persons immersed in the reconstructed virtual acoustic work environment may have a problem determining their orientation relative to the front/back direction specific for a given sound scene. In the course of recording the sound samples, the ambisonic microphone was placed

TABLE 3.1

Answers of Test Subjects to the Task Carried Out Based on Scenario A

Question 1: Street orientation	Perpendicular[a]	Parallel[a]	No idea
Number of answers	10	0	0
Question 2: Street orientation	In front	Behind	No idea
Number of answers	4	4	2
Question 3: Airplane location	High above	On the street level	No idea
Number of answers	8	0	2
Question 4: Airplane location	On the right	On the left	No idea
Number of answers	7	1	2

[a] To the direction the subjects were facing.

facing the street, however one half of all the subjects claimed that in the virtual acoustic environment, the street was behind them. There can be several causes of this problem—the most likely of which seem to be movement of the subject's head during the course of the test and the presence of too many reflected sounds recorded by the ambisonic microphone. The subjects were much more correct in determination of the position of individual objects making up the virtual acoustic environment in terms of left/right direction. This is clearly evidenced by responses to the question concerning orientation of the street relative to the listener and the task consisting in determination of the airplane flight path (in the course of recording the sound sample, the airplane flew approximately from the rear right to the front relative to the subject's position).

For the purpose of another test, Scenario B was developed in which a fragment of the recording from the virtual acoustic environment model of Scenario A was used. The sound scene duration was two minutes. The waveform of the sound signal in that scenario is shown in Figure 3.12.

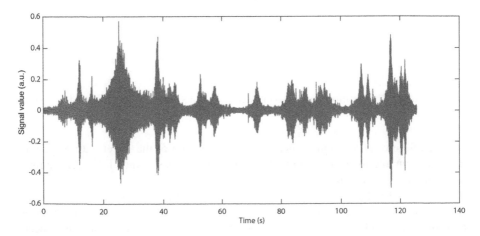

FIGURE 3.12   The waveform of the sound signal used in Scenario B.

TABLE 3.2

Answers of Tests Subjects to Questions of the Task Based on Scenario B

Question: The number of automobiles passing by from left to right was...	9	11	12	13
Number of answers	1	1	7	1

The subject's task consisted of mentally counting the number of motor vehicles passing by from left to right—made more challenging by instances when the recording included different vehicles moving in opposite directions at the same time. The scenario enabled to check the accuracy with which directional information about sound signals in uncomplicated situations is perceived in the specifically constructed virtual acoustic environment.

Table 3.2 summarizes responses from test subjects to the task based on Scenario B from which it follows that the majority of subjects were able to correctly identify the number of vehicles passing from their left to their right (12 cars). This information is of special value in view of the dynamic nature of situations involving passage of several automobiles at the same time. Results of the test confirm that the constructed virtual acoustic environment reproduced the actual conditions precisely enough to enable perception of objects moving in space and their direction of movement.

The next test was based on Scenario C in which the virtual acoustic environment modeled an office space dedicated to work of conceptual nature. The sound scene duration was 1 minute. A simplified schematic view of the office space in which the sound sample was recorded is shown in Figure 3.13. In the course of the recording, the microphone was placed in the room center very close to the workstations (about 1 m from each of the workstations). A printer was placed on the right at the angle of 45° relative to the scene front at the distance of about 1.5 m from the ambisonic

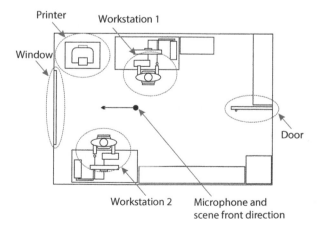

FIGURE 3.13 A simplified schematic view of the office space converted into the virtual work environment used in Scenario C.

microphone. In the course of recording, office work was conducted at both of the two workstations, the printer was switched on, and moreover, doors leading to a corridor were opened and shut several times to allow the sounds of people walking the hallway to be heard by the subjects.

Scenario C was created specifically to evaluate a subject's ability to gather a general impression and perceive the overall size of the virtual office setting from sound alone. The subjects' task was to listen intently to the presented sound scene and then, characterize the room and identify details of the pieces of equipment in use as well as the office activities carried out. Detailed questions asked of the subjects concerned position of workstations (including number thereof) and location of the printer and the doors (creaking when opened or closed). The subjects were also asked to express their opinions about the overall size of the office room in which they were virtually placed. Table 3.3 summarizes the answers of the subjects.

Answers to the asked questions show that most of the subjects managed to answer all of the questions correctly. In the majority of cases, the subjects were correct in assessing location of workstations and the printer in terms of left/right direction and probably made an association between the printer sound and the workstation position. As opposed to the preceding test, interpretation of door location turned out to be a less demanding task. The majority of subjects localized the door correctly at the back of the room. The issue of distance perception in virtual reality is a significant problem studied by many contemporary researchers, especially in the context of its dependence on the technique used for sound reproduction [Paquier et al. 2016]. Based on the realized virtual acoustic environment, the subjects were able to find that the room was rather small which is an evidence of the appropriateness of

TABLE 3.3

Answers of Subjects to Question of the Task Based on Scenario C

Question 1: Number of workstations	1–2	3–4	More than 4
Number of answers	7	3	0
Question 2: Workstation location	Left/right	In front/behind	No idea
Number of answers	7	1	2
Question 3: Printer location	Left/right	In front/behind	No idea
Number of answers	8	1	1
Question 4: Doors location	Behind	In front	No idea
Number of answers	6	1	3

reproduction of the actual environment in its virtual representation (the size of the office space can be assessed based on the distance to individual workstations and the reverberation time, therefore the information was correctly reconstructed in the virtual environment).

For the purpose of the next test, acoustic Scenario D was prepared constituting a virtual acoustic model reproducing the street traffic occurring at a measurement point located on a busy main road in a highly urbanized city area. As part of the test, the subjects were asked to identify the direction from which the test signal in the form of a bicycle bell was coming. The whole sound scene lasted two minutes, in the course of which a sequence of 24 bell signals were generated and separated from each other by 3-second intervals. To generate the bell signal, the test stimuli playback system described in the preceding section was used. The direction from which the bell signal was coming varied randomly, and at the same time, in the whole sequence for each direction, the bell signal was replayed three times. Figure 3.14 shows the structure of bell ring sequences in the Adobe Audition application used to manage the stimuli playback system.

Figure 3.15 shows the waveform of the bell ring signal emitted in the course of the test. The sound pressure level of the created street traffic acoustic image varied in time due to continuous motion of vehicles in the reconstructed scene. Figure 3.16 presents the waveform corresponding to the road traffic sound signal played back in the test scenario with marks identifying time intervals in which the

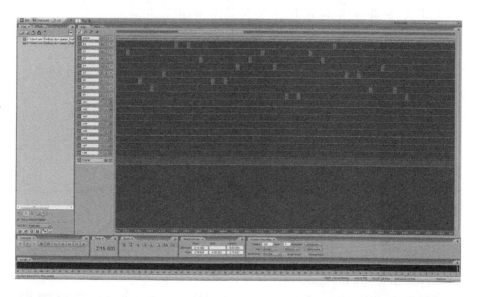

FIGURE 3.14 A view of program window used to manage the stimuli playback system with programmed sequence of bicycle bell sound signals.

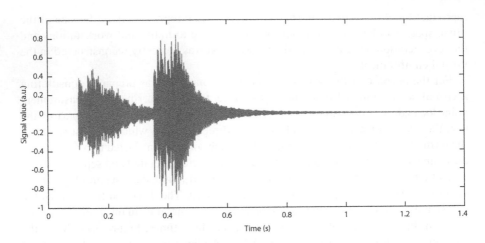

FIGURE 3.15   The bicycle bell ring signal waveform.

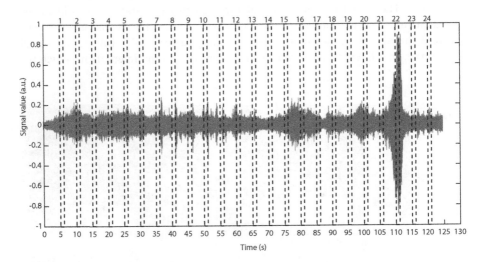

FIGURE 3.16   The street traffic signal waveform with marks identifying moments at which the bicycle bell ring signals were emitted.

bell signal was emitted. Note that signal No. 22 was played back at the moment of a significant increase of the road traffic sound signal amplitude due to the passage of a motorcycle as shown in Figure 3.17.

Creation of that specific acoustic scenario enabled researchers to focus on the directivity of hearing in the subjects. The ambisonic technique can be employed to create the virtual acoustic environment of a busy city street, whereas test signals are realized with the use of a separate monophonic system, therefore the test results can be considered objective. In Scenario D, in order to familiarize the subjects with possible directions from which emission of the bell signals might occur, the test was

FIGURE 3.17 A view of the street crossing at which the sound sample was recorded with the identified sound source (a motorcycle) marked with a circle.

preceded with a trial playback of bell ring signal sequences without simultaneous reproduction of the street traffic sound image. In the course of performing the task, the subjects pointed their hands at the direction from which, in their opinion, the bell ring sound was coming. A dedicated form was developed to record answers of individual subjects. Table 3.4 summarizes answers of test participants to the task given them in Scenario D.

The task tested the directivity of hearing a bicycle bell sound immersed in noise generated by urban street traffic. In Table 3.4, directions from which the bell signal was actually coming are represented numerically in the column "Correct answer" with number 1 corresponding to the direction right ahead and successive integers representing directions separated by the angle of 45° clockwise. It follows from the recorded answers that in conditions characterized with street traffic noise, three directions were recognized with the highest accuracy, namely "in front" (direction 1), "on the right" (3), and "on the left" (7), for which from 60% to 90% correct answers were obtained. Identification of the "behind" and intermediate directions (especially those in the rear half-plane) turned out to be less successful. Two bell signals, numbered 20 and 22, remained unrecognized by most of the subjects. This is probably a result of the bell ring sound being masked by the street traffic noise. In case of signal No. 20, in the background noise it is possible to identify a car drive, train or brake system squeal as well as a car horn sound; further, signal No. 22 is drowned out by the passage of a noisy motorcycle which illustrates the effect of particularly dynamic situations from real life impacting on identification of direction from which sound signals come to the listener. Results of the test indicate that the virtual acoustic environment created in this way can be used in studies on directivity of hearing in acoustically diverse conditions with emphasis put on appropriateness of created sound scenes and their effect on results of directivity assessment tests.

TABLE 3.4

Results of the Test Concerning Identification of the Direction from Which the Bell Ring Signal Was Coming (Scenario D)

Signal No.	Correct Answer	Subject ID										% Correct
		S1	S2	S3	S4	S5	S6	S7	S8	S9	S10	
1	2	2	2	5	3	3	3	3	3	3	5	20%
2	6	6	8	7	8	7	7	8	7	6	7	20%
3	7	6	7	7	7	7	7	7	7	7	7	90%
4	5	5	4	1	5	5	6	1	4	8	0	30%
5	1	5	1	1	1	1	1	2	1	1	1	80%
6	1	1	1	1	8	5	8	1	1	1	1	70%
7	3	3	3	3	0	3	3	3	3	4	3	80%
8	6	6	6	6	8	6	7	7	6	7	7	50%
9	6	6	8	7	8	6	8	7	6	7	7	30%
10	4	4	4	3	2	4	4	3	4	4	3	60%
11	3	3	3	3	3	2	4	3	3	3	3	80%
12	2	2	2	2	2	2	3	3	3	2	2	70%
13	4	4	3	4	3	4	5	3	3	4	1	40%
14	8	7	7	7	8	7	7	8	7	8	7	30%
15	8	7	8	7	8	7	8	7	7	8	8	50%
16	2	0	2	4	2	4	3	3	3	2	3	30%
17	3	3	3	3	3	3	4	3	3	3	2	80%
18	1	8	8	1	1	5	1	1	5	1	1	60%
19	5	5	5	8	6	5	0	2	5	8	1	40%
20	5	0	4	0	0	0	0	0	0	8	0	0%
21	7	7	8	7	8	7	8	7	7	7	7	70%
22	4	4	0	0	0	3	0	0	0	4	0	20%
23	7	7	8	6	8	7	7	7	7	7	7	70%
24	8	8	8	8	0	7	6	8	6	8	8	60%
% Correct per subject		71%	58%	50%	42%	54%	29%	42%	50%	71%	46%	

In the last test, Scenario E was used which also included a sound scene based on a recording made in a busy conurbation. The scene was recorded in front of a zebra crossing—a pedestrian crossing marked by alternating light and dark stripes painted parallel to the flow of traffic—at an intersection of busy streets equipped with an acoustic signaling system. The duration of the sound scene was five minutes. The subjects' task was to signal the moment at which it was possible to cross the street safely. The guess was communicated to the test conducting person by the subject by raising their hand. The reconstructed virtual acoustic environment model was characterized with significant complexity. Figure 3.18 shows the waveform of the sound signal used in that scenario with frames marking the time intervals in which the green traffic light for the pedestrian crossing was on. Moreover, individual sound sources are assigned to selected characteristic sound events in the replayed scene.

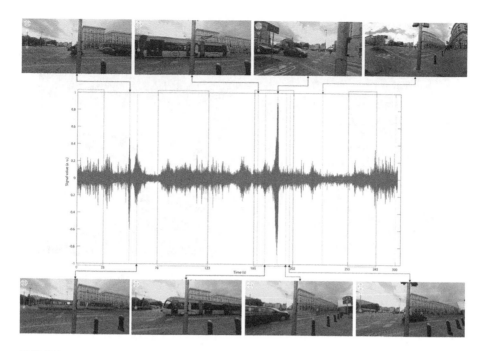

FIGURE 3.18 The waveform of the street traffic sound signal used in Scenario E.

This virtual acoustic environment model encompassed a multitude of specific situations, including, among other things, a streetcar passing by across the pedestrian crossing in the course of which the crossing light turned green and passage of motor cars across the zebra in the course of conditional right turn (while the pedestrians' green light was on). The noise generated by cars going (parallel to the crossing at which the recording was registered) over bumps of the tram line running along the street could be also perceived as mysterious and affecting feelings of test subjects in a specific way. On a pedestrian crossing perpendicular to that in front of which the sound signal was recorded, there was another signaling sound system for the visually impaired which could also affect the interpretation of the traffic situation in the subjects. It should be also borne in mind that one of the streets (see Figure 3.19) at the recording site was a one-way road for the motor traffic (in the recording, motor vehicles passed by from left to right), while trams were going along the street in both directions which might be the cause of some disorientation in the subjects.

Table 3.5 summarizes answers of test subjects to the task based on the scenario in which they were asked to indicate the time intervals in which the green traffic light was on in the recording. The actual time intervals safe for crossing the street (green pedestrians' streetlight on) correspond to the following segments of the recording (in seconds): 0–25, 76–123, 165–202, and 255–282.

The city crossroad traffic used to realize the virtual acoustic environment model was characterized with high complexity and intensity. The visually impaired subjects described the task as a very difficult one. Despite existing and operational sound signaling systems, three subjects confessed that they would be too scared to

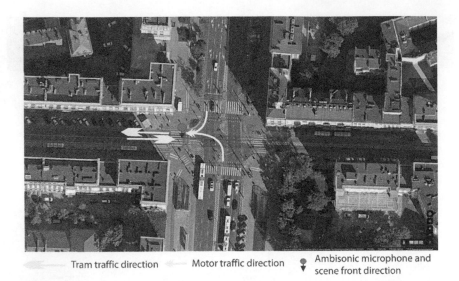

Tram traffic direction   Motor traffic direction   Ambisonic microphone and scene front direction

FIGURE 3.19 A view of the actual measurement site used to create Scenario E. (Map source: Google Maps.)

TABLE 3.5
Answers of Test Subjects to the Task Based on Scenario E

Subject ID	The Moments/Intervals Indicated as Safe for Crossing the Street (recording time in seconds)
S1	—
S2	5; 78; 168; 257
S3	9–14; 35; 83–91; 103–117; 170–180; 262–268; 285
S4	4–23; 72–126; 168–248; 269–300
S5	80; 169; 257
S6	37–43; 67–79; 138–148; 215, 219; 237–243
S7	30–41; 81–107; 172–187; 259
S8	30–56; 71–80; 143–151; 257–263
S9	35; 84; 191
S10	70; 144; 229; 259

use the crossing on their own. The scene began with the green traffic light switched on for 25 seconds. Decision on starting the scene with that specific sound was aimed at checking whether the subjects would be able to recognize the situation right away—from the very start of the exercise. Three subjects made their decision to cross the street with the green light switched on for the first time (i.e., at the beginning of the scene). All three subjects were visually impaired persons which may indicate a better feel for the scene among that group of subjects compared to results achieved by persons with good vision. It is also worth emphasizing that two of the gravely visually impaired persons were able to identify the safe cross-ing time intervals with a 100% hit results. In case of that recording, these subjects had no problems with correct interpretation of the environment in which they were virtually placed. In case of switching the green traffic light for the second, third, and fourth time, the percentage of correct answers was 70%, 60%, and 70%, respec-tively. One of the visually impaired persons was unable to indicate any safe moment for crossing the street. Two visually impaired subjects claimed that their decisions were based on helpful, to some extent, audible signals coming from the neighboring pedestrian crossing. One of the subjects was in error in every case where their hand was raised to indicate the moment which is safe for crossing the street. From fur-ther analysis of the moments indicated by that subject as suitable for safe crossing, it can be noticed that this person planned to cross the roadway based exclusively on recognition of the background noise (allegedly safe moments for crossing the street were those associated with lower traffic noise levels). This subject entirely ignored the acoustic signaling system. Indications from other subjects included dif-ferent numbers of misjudgments, from isolated errors to several cases where wrong answers predominated. Results of tests realized with the use of Scenario E indicate that for visually impaired persons, going through hearing training focusing on the recognition of sounds typical for pedestrian crossings—especially in specific loca-tions (zebra crossings used most frequently by such persons)—can be a matter of great significance. This virtual environment scenario correctly and with adequate accuracy reproduces the actual acoustic environment and can be used to provide training in the area of spatial imagination skills.

3.5 SUMMARY

Virtual acoustic environments play an important role in contemporary research on the perception of sound signals and constitute a basis for further research on assess-ment of hearing directivity in employees in different work environments. To sum up the presented research results it can be concluded that the laboratory setup for sound environment auralization with the use of ambisonic techniques enabled the test subjects, in the majority of cases, to perceive the sound correctly. The situations created in laboratory conditions were characterized with high level of realism of reproduction of actual sound environments. Therefore, the developed setup can be used for planning and carrying out both scientific experiments and training activi-ties concerning sound perception and spatial imagination, especially for visually impaired persons.

In the case of developing one's own test scenarios for specific research purposes it is worthwhile to consider the option of using recordings available in databases in order to compare the results of own studies with conclusions from research projects carried out in other laboratories [Weisser et al. 2019] and make use of the guidelines concerning creation of virtual audio environments [Lindau et al. 2014]. The course of conduct adopted in the presented test procedures enabled our researchers to assess perception of test signals and correctness of subjective aural impressions in the sub-jects. Results of the test confirmed usefulness of the setup for research oriented at similar objectives.

REFERENCES

Assenmacher, I., T. Kuhlen, T. Lents, and M. Vorlander. 2004. Integrating real-time binaural acoustics into VR applications. *Proceedings of the Tenth Eurographics Conference on Virtual Environments*, June 8–9, 2004, Grenoble, France, pp. 129–136.

Bălan, O., A. Moldoveanu, and F. Moldoveanu. 2015. Multimodal perceptual training for improving spatial auditory performance in blind and sighted listeners. *Arch Acoust* 40(4):491–502. DOI:10.1515/aoa-2015-0049.

Bertet, S., J. Daniel, E. Parizet, and O. Warusfel. 2013. Investigation on localisation accuracy for first and higher order smbisonics reproduced sound sources. *Acta Acust United Acust* 99(4):642–657. DOI:10.3813/AAA.918643.

Betlehem, T., C. Anderson, and M. A. Poletti. 2010. A directional loudspeaker array for surround sound in reverberant rooms. *Proceedings of the 20th International Congress on Acoustics 2010, ICA 2010, Incorporating 2010 Annual Conference of the Australian Acoustical Society*, August 23–27, 2010, Sydney, Australia, Vol. 2, pp. 918–923.

Carlsson, K. 2004. Objective localisation measures in ambisonic surround–sound. Master Thesis in Music Technology. Stockholm: Royal Institute of Technology. www.speech. kth.se/prod/publications/files/1663.pdf (accessed January 28, 2020).

Clapp, S., J. Braasch, A. Guthrie, and N. Xiang. 2014. Evaluating the accuracy of the ambisonic reproduction of measured soundfields. *Proceedings of the EAA Joint Symposium on Auralization and Ambisonics*, April 3–5, 2014, Berlin, Germany. https://depositonce. tu-berlin.de/bitstream/11303/185/1/27.pdf (accessed January 28, 2020).

Deldjoo, Y., and R. E. Atani. 2016. A low-cost infrared-optical head tracking solution for virtual 3D audio environment using the Nintendo Wii-remote. *Entertain Comput* 12:9–27. DOI:10.1016/j.entcom.2015.10.005.

Dunai L., I. Lengua, G. Peris-Fajarnes, and F. Brusola. 2015. Virtual sound localization by blind people. *Arch Acoust* 40(4):561–567. DOI:10.1515/aoa-2015-0055.

Echevarria Sanchez, G. M., T. V. Renterghem, K. Sun, and B. De Coensel. 2017. Using virtual reality for assessing the role of noise in the audiovisual design of an urban public space. *Landsc Urban Plan* 167:98–107. DOI:10.1016/j.landurbplan.2017.05.018.

EN ISO 7731:2008. 2008. Ergonomics—Danger signals for public and work areas—Auditory danger signals. Brussels, Belgium: European Committee for Standardization.

Epain, N., C. T. Jin, and F. Zotter. 2014. Ambisonic decoding with constant angular spread. *Acta Acust United Acust* 100(5):928–936(9). DOI:10.3813/AAA.918772.

Gerzon, M. A. 1973. With-height sound reproduction. *J Audio Eng Soc* 21(1):2–10.

Gerzon, M. A. 1974. Surround-sound psychoacoustics. *Wireless World*. www.audiosignal.co.uk/ Resources/Surround_sound_psychoacoustics_A4.pdf (accessed January 28, 2020).

Hammershoi, D., and H. Møller. 2002. Methods for binaural recording and reproduction. *Acta Acust United Acust* 88(3):303–311. https://vbn.aau.dk/ws/portalfiles/portal/ 227975991/2002_Hammersh_i_and_M_ller_AA.pdf (accessed January 28, 2020).

Harley, R. 2010. *The Complete Guide to High-End Audio*, 4th ed. Chicago: Acapella Publishing.

Heller, A. J., E. Benjamin, and R. Lee. 2010. Design of ambisonic decoders for irregular arrays of loudspeakers by non-linear optimization. *Proceedings of the 129th Convention of the Audio Engineering Society*, 4–7 November 2010, San Francisco. www.ai.sri.com/ ajh/ambisonics/BLaH4.pdf (accessed January 28, 2020).

Heller, A. J., E. M. Benjamin, and R. Lee. 2012. A toolkit for the design of ambisonic decoders. *Proceedings of the Linux Audio Conference 2012*, 12–15 April 2012. CA. http://lac. linuxaudio.org/2012/papers/18.pdf (accessed January 28, 2020).

Heller, A. J., R. Lee, and E. M. Benjamin. 2008. Is my decoder ambisonic? *Proceedings of the 125th Convention of the Audio Engineering Society*, 2–5 October 2008. San Francisco.

Hojan, E., M. Jakubowski, A. Talukader et al. 2012. A new method of teaching spatial orientation to the blind. *Acta Phys Pol A* 121(1–A):A5–A8. DOI:10.12693/APhysPolA.121.A-5.

Lindau, A., Erbes, V., Lepa, S., Maempel, H.-J., Brinkmann, F., and Weinzierl, S. 2014. A Spatial audio quality inventory (SAQI). *Acta Acust United Acust* 100(5):984–994(11). DOI:10.3813/AAA.918778.

Marentakis, G., F. Zotter, and M. Frank. 2014. Vector-base and ambisonic amplitude panning: A comparison using pop, classical, and contemporary spatial music. *Acta Acust United Acust* 100(5):945–955(11). DOI:10.3813/AAA.918774.

Mlynski, R., and E. Kozlowski. 2019. Localization of vehicle back-up alarms by users of level-dependent hearing protectors under industrial noise conditions generated at a forge. *Int J Environ Res Public Health* 16(3):e394. DOI:10.3390/ijerph16030394.

Moore, D., and J. P. Wakefield. 2008. The design of ambisonic decoders for the 5.1 layout with even performance characteristics. *Proceedings of the 124th Audio Engineering Society Convention*, 17–20 May 2008, Amsterdam, the Netherlands.

Morzynski, L., D. Pleban, G. Szczepanski, and R. Mlynski. 2017. Laboratory for testing of sound perception in virtual acoustic working environment. *48° Congreso Español de Acústica; Encuentro Ibérico de Acústica; European Symposium on Underwater Acoustics Applications; European Symposium on Sustainable Building Acoustics*, 3–6 October 2017, A Coruña, Spain, pp. 401–406.

Paquier M., N. Côté, F. Devillers, and V. Koehl. 2016. Interaction between auditory and visual perceptions on distance estimations in a virtual environment. *Appl Acoust* 105:186–199. DOI:10.1016/j.apacoust.2015.12.014.

Pelzer, S., B. Masiero, and M. Vorlander. 2014. 3D reproduction of room auralizations by combining intensity panning, crosstalk cancellation and ambisonics. *Proceedings of the EAA Joint Symposium on Auralization and Ambisonics*, 3–5 April 2014, Berlin, Germany. https://core.ac.uk/download/pdf/57699739.pdf (accessed January 28, 2020).

Pulkki, V. 1997. Virtual sound source positioning using vector base amplitude panning. *J Audio Eng Soc* 45(6):456–465.

Rabenstein, R., S. Spors, and J. Ahrens. 2014. Sound field synthesis. *Signal Process* 4:915–979.

Recommendation ITU-R BS.2094-1. 2017. Common definitions for the audio definition model, Geneva. www.itu.int/rec/R-REC-BS.2094-1-201706-I/en (accessed May 29, 2020).

Satongar, D., C. Dunn, Y. Lam, and F. Li. 2013. *Localisation Performance of Higher-Order Ambisonics for Off-Centre Listening*. British Broadcasting Corporation Reasearch & Development. www.bbc.co.uk/rd/publications/whitepaper254 (accessed January 28, 2020).

Sawicki, J., and W. Mickiewicz. 2006. Influence of acoustical properties of a listening room on the consistency between sound fields reproduced using standard stereo technique and headphone technique with HRTF processing. *Arch Acoust* 31:4(S):343–348.

Stitt, P., S. Bertet, and M. V. Walstijn. 2014. Off-centre localisation performance of ambisonics and HOA for large and small loudspeaker array radii. *Acta Acust United Acust* 100(5):937–944.

Szczepański, G., L. Morzyński, D. Pleban, and R. Młyński. 2018. Stanowisko badawcze CIOP-PIB do badań percepcji dźwięku przestrzennego z zastosowaniem techniki ambisonicznej. [CIOP-PIB test stand for studies on spatial sound perception using ambisonics]. *Bezpieczeństwo Pracy – Nauka i Praktyka* 10:24–27. DOI:10.5604/01.3001.0012.6477.

Vennerod, J. 2014. Binaural reproduction of higher order ambisonics – A real-time implementation and perceptual improvements. Master Thesis. NTNU – Trondheim: Norwegian University of Science and Technology. https://ntnuopen.ntnu.no/ntnu-xmlui/handle/11250/2371408?show=full (accessed January 28, 2020)

Weisser, A., J. M. Buchholz, C. Oreinos et al. 2019. The ambisonic recordings of typical environments (ARTE) database. *Acta Acust United Acust* 105(4):695–713(19). DOI:10.3813/AAA.919349.

Williams, M. 1987. Unified theory of microphone systems for stereophonic sound recording. *Proceedings of the 82nd Audio Engineering Society Convention*, 10–13 March 1987, London: Great Britain – AES Preprint 2466.

Yao, S. N. 2017. Equalization in ambisonics. *Appl Acoust* 139:129–139.

Zhang, W., P. N. Samarasinghe, H. Chen, and T. D. Abhayapala. 2017. Surround by sound: A review of spatial audio recording and reproduction. *Appl Sci-Basel* 7(5):532. DOI:10.3390/app7050532.

4 Wireless Sensor Networks

Leszek Morzyński

CONTENTS

4.1 INTRODUCTION TO WIRELESS SENSOR NETWORKS (WSN) AND THE INTERNET OF THINGS (IoT)

The simplest way to define the wireless sensor network [Raghavendra et al. 2006; Yick et al. 2008; Dargie and Poellabauer 2010; Yang 2014] is to describe it as a group of sensors deployed in a portion of physical space and communicating with each other by means of a wireless communication method for the purpose of collecting and processing data on condition of physical systems, processes, or environments. In this definition, a sensor should not be considered as just a transducer converting a physical quantity into an electric signal but rather a specialized module comprised of such a transducer, circuits for processing and conditioning of the measuring signal (in particular microcontrollers), and a radio communication module. Such sensor modules are known as *motes*, especially in North America. Many sensor networks

include also actuators in their structure and are therefore capable of not only collecting data concerning physical systems but also actively controlling their parameters (influencing the operation of a system or condition of an environment, or even
controlling a process). Such a network is therefore used to transmit not only the
measurement data but also control commands. As far as sensor networks comprising
both sensors and actuators are concerned, two kinds of nomenclature are used in the
literature. Some authors prefer using a wider definition of wireless sensor networks
comprising both sensors and actuators (cf. e.g., [Yang 2014]); in other publications,
such a network is called the wireless sensor and actuator network [Vergone et al.
2008]. In the present chapter, the first of option is used—it is understood that a
"wireless sensor network" comprises both sensors and actuators.

Devices making up a WSN are called *network nodes* and the tasks they are
assigned in the network may be different. Apart from the above-mentioned sensors
and actuators, a node may function as, for example, a network coordinator or a router.
Each network node is connected with at least one other node. Each WSN also has at
least one so-called *sink node* to which data from sensor nodes are conveyed. The sink
node is the point of network interface with external systems which process and analyze the data collected by the network and an interface with the end user of such data.

A landmark step in the development of wireless sensor networks consisted in connecting them to the Internet. Such connection opens a wide range of areas in which
WSNs can be employed and broadens the range of benefits which can be drawn from
such employment. The end user of a WSN may access it (to e.g. acquire data, control
functioning, and/or issue commands) any time and from any place where access to
Internet is provided. A WSN is also capable of sending data from sensor modules
to the computing cloud for processing (in case of a shortage of resources such as
computing power or storage capacity). What is perhaps most important is that the
devices making up a WSN can communicate directly with other devices connected
to Internet (without participation of a human), exchanging data and/or commands.
Such direct information exchange between devices is called machine-to-machine
(M2M) or, more rarely, thing-to-thing (T2T) communication. Connecting WSNs to
the Internet marked the emergence of the Internet of Things (IoT) or the Internet in
which machines communicate with machines [Ashton 2009; Chabanne et al. 2011;
Miorandi et al. 2012; Delicato et al. 2013; Gubbi et al. 2013; Galio and Lo Re 2014;
Karimi and Atkinson 2014]. The term was used for the first time in 1999 by Kevin
Ashton in his presentation prepared for the Procter & Gamble company [Ashton
2009]. Development of WSN and IoT has paved the way for changes that have been
termed the "fourth industrial revolution" or Industry 4.0 (the term was used for the
first time in 2011 in the German government's strategy concerning new technologies
[Kagermann et al. 2011]). The concept of Industry 4.0 assumes integration of manufacturing systems and sensor networks into structures offering the possibility to
decentralize decision-making processes [Brettel et al. 2014]. This allows creation of
manufacturing environments optimized with regard to the use of available resources
and capable of responding dynamically to varying market needs. Currently, WSNs
and IoT are widely used in, among other things [Atzori et al. 2010; Miorandi et al.
2012; Gubbi et al. 2013; Karimi and Atkinson 2014; Yang 2014; Wortmann and
Fluchter 2015]:

- automation and monitoring of industrial processes,
- managing traffic and controlling lighting systems in urban areas (the so-called smart cities),
- managing environment in residential buildings including control of, among other things, lighting, heating, air conditioning and ventilation (the so-called smart homes),
- monitoring patients' state of health in real time and diagnosing diseases,
- monitoring of the condition of natural environment, and
- monitoring and managing vegetable farming in agriculture.

These examples of the application of WSNs and IoT show that solutions of this type can, and obviously should, be used for the purpose of shaping a safe and human-friendly work environment. Solutions similar to those employed to monitor conditions prevailing in natural environments, watch the state of health in a patient, or manage environment in a building can be also used to construct wireless sensor networks which monitor hazards existing in work establishments and intelligently affect behaviors and actions of employees and employers, thus avoiding various hazards of which noise is one of the most common occurrences. The present chapter includes a discussion of fundamental issues concerning design and realization of wireless sensor networks, followed by presentation of model WSN solutions supporting reduction of workers' exposure to noise in their work environment.

4.2 CHOICE OF STRUCTURE AND PARAMETERS FOR A WNS IN VIEW OF ITS INTENDED APPLICATION

Realization of a sensor network designed to carry out a specific task involves the necessity of choosing implementation methods appropriate for the given application. This should be understood as the need to decide what structure the network will be given and which wireless data transmission standard will be used, with the two issues being interrelated in many cases. Subsequent steps include the selection of suitable technical and structural solutions which can be employed to construct the network.

The main factors necessary to be taken into account when choosing the network structure and communication standard are:

- the network size understood as the number of devices making up the network,
- the size of the area in which the network operates,
- the maximum transmission range required for the network,
- the required data transfer rate,
- the limitations concerning overall dimensions and electric energy consumption characterizing the modules making up the network, and
- the tasks to be carried out by the network (for instance, collection of data from sensor modules only or together with identification of location of network elements).

Depending on the size of the area they are designed to cover, wireless networks can be categorized as follows:

- Wireless Local Area Network (WLAN)—a network covering a workplace, a building or a portion of such a workplace, within a range of several hundred meters,
- Wireless Personal Area Network (WPAN)—a network with a small range of up to several dozen meters, capable of covering a specific marked-off area within a work establishment site, or
- Body Area Network (BAN), known also as Wireless Body Area Network (WBAN) or Body Sensor Network (BSN)—a network in direct vicinity of a human body, with the range limited in most cases to wearable devices.

Small-sized networks may communicate with higher-order networks covering larger areas with the use of so-called *gateways*.

Basic structures of a wireless sensor network are presented in Figure 4.1.

In networks with the *star structure*, all network elements (marked S in Figure 4.1) connect directly to one central network element (C/G) functioning as the network coordinator and at the same time the network gateway (sink node). The main advantage of networks of this type is the ease of realization of their embodiments. In networks with the *mesh structure*, each element of the network may play the role of the network router and communicate with any other element of the network. A network of this type can operate over larger areas and is more resistant to signal transmission interference and network structure damage (any communication from point A to point B may be transmitted over different paths, depending on the current situation in the network). The *tree structure* is a topology between the star and the mesh structures which, on one hand, imposes some additional limitations concerning data transmission paths compared to the mesh topology, but on the other, algorithms for finding optimal data transmission paths are simpler in networks based on the tree structure.

Among the most commonly known and used wireless communication standards which can be applied in wireless sensor networks, one should name:

- Wi-Fi, based on IEEE 802.11a/b/g/n standards, a standard solution for up-to-date local computer networks, characterized by high data transfer rate;
- Bluetooth and Bluetooth Low Energy (Bluetooth LE, BLE, Bluetooth Smart)—a family of protocols based initially on IEEE 802.15.1 standard

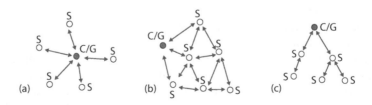

FIGURE 4.1 Basic network structures (topologies): (a) star; (b) mesh; and (c) tree.

and currently developed by Bluetooth Special Interest Group (SIG), used mainly for short-distance communication;

- ZigBee, based on IEEE 802.15.4 standard, developed mainly with the intent to be used in WPAN sensor networks as an easier to implement and less expensive alternative to Bluetooth protocol, characterized with low data transfer rate and low power consumption level;
- Thread, based on 6LoWPAN (IPv6 over Low-Power Wireless Personal Area Networks) standard derived from IEEE 802.15.4 standard, developed with the intent to be used in sensor WLANs, distinctive feature of which is implementation of IPv6 Internet Protocol;
- Z-Wave, developed by Danish company Zensys and currently supervised by Z-Wave Aliance, designed mainly for household automation applications; and
- IEEE 802.15.6, developed specifically for BAN applications, characterized with extremely low power demand and high privacy protection level, currently still not widely utilized.

Table 4.1 presents a comparison of the selected wireless communication standards. For each item, the following are given:

- available network topology options: peer-to-peer (P2P), star, mesh, or tree,
- relative power demand,
- data transfer rate, and
- typical range (may vary depending on the adopted class of the solution, transmitter power and structure, whereas the maximum power of the transmitter and thus also its range may be subject to limitations following from specific local regulations).

In any project aimed at realization of a specific wireless sensor network it is necessary to take into account, besides the above-listed wireless communication standards, the RFID (Radio-Frequency Identification) technique which is needed to identify and track objects. RFID is based on radio communication between two devices (the tag and the reader). The RFID technique is defined by a number of standards concerning, among other things, different areas of its application. In practical embodiments of the technique, passive or active tags may be used. Passive tags do

TABLE 4.1
A Comparison of Properties of Different Wireless Communication Standards

	Bluetooth/ Bluetooth LE	Wi-Fi	ZigBee	Z-Wave	Thread
Topology	Star	Star	Mesh, star, tree	Mesh	Mesh, star
Power consumption	Very low to low	Low to high	Very low	Very low	Very low
Data transfer rate	125 kb/s–50 Mb/s	11 Mb/s–10 Gb/s	40–250 kb/s	9.4–200 kb/s	250 kb/s
Range (m)	10–140	30–140	10–300	30–100	800

not have their own power supply source but when placed in an electromagnetic field generated by a reader, they store the electric energy received via their antenna circuit in a capacitor integrated into the tag structure. Once a sufficient amount of energy is collected, the tag sends a reply signal containing its code. Active tags use their own power source to send out their radio information messages. A reader present within the range can read such a message and identify the tagged object. Bluetooth beacons are often used as active tags. That is a tagging solution in which, for the purpose of application of RFID technique, Bluetooth protocol is used (each beacon sends an advertising signal defining its unique identifier which can be received by any Bluetooth "reader").

The functionality of a sensor network supporting reduction of workers' exposure to noise in the work environment can be extended by application of solutions from the area of IoT. Exchange of data on the Internet requires that the Internet Protocol (IP) is used in the network layer of an OSI model (ISO Open Systems Interconnection Reference Model). Devices exchanging data on the Internet are identified by an IP address assigned to them. To secure communication between devices on the Internet, transport layer protocols such as TCP (Transmission Control Protocol) or UDP (User Data Protocol) or application layer protocols such as HTTP (HyperText Transfer Protocol) are also used. The Wi-Fi standard (IEEE 802.11) defines the two lowest layers of OSI model (the physical layer and data link layer). As Wi-Fi technology is used to create internet networks, Wi-Fi standard implementing devices also have also IP management implemented in the network layer. Such devices may be easily connected to the Internet. The situation is different in the case of Bluetooth and ZigBee protocols which define higher layers of the OSI model, including a network layer protocol different from the IP—in Bluetooth protocol, it is possible to transmit packages compatible with IP protocol with the use of encapsulation (however, such solutions have their own limitations), whereas some development versions of ZigBee implement the IP protocol as their standard and are called ZigBee IP. Connection of a WSN with the Internet depends therefore on the communication standard used by the network. If a WSN uses a standard implementing the IP in the network layer then the network can be connected with the Internet by means of a classic (Internet) gateway. In cases when the communication standard used in WSN does not implement the IP, connection with the Internet must be realized by means of a gateway in which software is installed enabling translation of protocols (in particular device addresses). In view of the rapid development of the IoT, gateways of that specific type (IoT Gateways) are more and more commonly offered by computer hardware manufacturers. In the simplest case, data exchange between a WSN and the Internet may be effected by means of an intermediary server (proxy). Such intermediary servers include installed software that manages reception of data from WSN and stores that data in a database, as well as software that enables the stored information to be available via Internet.

The intended application of any sensor network is an important consideration during design (and prior to construction) of the network. For example, if there is the need to locate moving objects or those holding or wearing moveable elements, those requirements must influence the structure of the network. Such an object could be, for instance, a worker for whom the network is configured to assess his exposure to noise or alert him or her to such hazards. That network function implies the necessity to identify the worker's position relative to noise hazard areas. In the following, two options are presented to accomplish this task together with sensor network structural options most suitable to accomplish these goals.

The first option consists in using RFID techniques in combination with division of the network operation area into zones. The RFID techniques are widely used in access control systems. Generally, the entrances to restricted access areas are controlled/equipped with a RFID reader module, whereas the person requiring access is equipped with a tag (the opposite solution is possible, depending on structure of the actual network embodiment). The access control system will permit entrance to a specific area (e.g., by opening a gate or door) for a person bearing the proper tag and can also collect information about individuals who have entered a given zone using information coded in their tags. Such solutions can be therefore used to evaluate actual position of workers relative to hazard zones. It involves division of the work enterprise site area into zones which separate out the hazard occurrence areas and establishment of permissible entrances to such zones equipped with RFID readers. Thanks to application of RFID techniques, any instance of a worker's entrance to a hazard zone would be noticed in the sensor network and a warning message would be sent to the worker. Such solution would perform most effectively in cases where it is easy to demarcate hazard zones, especially when such zones encompass the whole area of a workroom. An example structure of such network is shown in Figure 4.2. In the presently considered example, active RFID tags in the form of Bluetooth beacons were used and Wi-Fi was adopted as the wireless communication standard.

In the case shown in Figure 4.2, the workplace is divided into three areas (rooms) which can be accessed through specifically designated passages (doors, gates). The whole of each of the areas is treated by the network as a hazard zone (although the actual hazard area may not cover the whole of the zone). Such network must

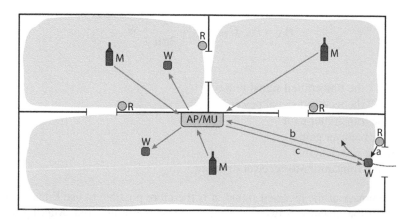

FIGURE 4.2 Structure and the principle of operation of a sensor network with hazard zone access control (M—meter/sensor; W—wearable device; R—RFID device—a Bluetooth beacon; AP/MU—access point connected with main unit; a, b, c—communication establishing/data transmission sequence; black arrow—Bluetooth LE data transmission direction; gray arrows—directions of Wi-Fi data transmission; gray areas—hazard zones).

include a main unit (which will be typically a PC-class computer) collecting and processing data transmitted in the network. The main unit (MU) communicates with wireless devices present in the network via an access point (AP). In Figure 4.2, the access point connected with the main unit is represented by the AP/MU block. The person whose position relative to hazard zones is examined must obviously have on him/her a sensor or actuator module being an element of the sensor network. Such module may have the form of a wearable device.

When a person equipped with a wearable device bearing a network module enters any of the demarcated zones, the wearable device records the fact by reading a signal from RFID device installed at the entrance (transmission marked with symbol "a" in Figure 4.2). Next, the wearable device sends information to the main unit about the fact of that particular wearable device being found in a specific hazard zone (transmission marked with symbol "b" in Figure 4.2). In return, the main unit advises the wearable device of hazards existing in the zone (transmission marked with symbol "c" in Figure 4.2). Measuring devices functioning in the network transmit the data on existing hazards to the network main unit. The main unit collects the data, processes it and, if the need arises, sends messages to other wearable devices present in the zone.

Another method for locating the position of a worker is based on evaluation of the distance from the radio transmitter with the use of the received signal power value and the triangulation method [Yang 2014]. This method allows to determine the absolute position of a worker in a predefined coordinate system within the area of operation of the sensor network.

The distance evaluation method is based on the received signal power value and uses the fact that the received signal power is a function of the distance to the transmitter. In open space, the relationship between the transmitted signal power and the received signal power has the form of the Friis transmission formula:

$$P_{RX} = P_{TX} \cdot G_{TX} \cdot G_{RX} \cdot \left(\frac{\lambda}{2\pi d} \right)^2 W \qquad (4.1)$$

where:
P_{TX} (W): the transmitted signal power
P_{RX} (W): the received signal power
G_{TX}: the transmitter gain
G_{RX}: the receiver gain
λ (m): the wavelength
d (m): the transmitter-to-receiver distance

Given the power of the received signal, it is therefore possible to use Equation (4.1) to calculate the distance between the receiver and the transmitter. Making use of this relationship in practice became easier thanks to development of wireless data transmission technology. In wireless data transmission networks, the received signal power has an important effect on the process of establishing connections between individual network devices and on the data transmission quality. The physical layer of network protocols was extended by introducing the ability to measure the received signal power. In IEEE 802.11 standard on which Wi-Fi technology is based, the concept of RSSI (Received Signal Strength Indicator) was introduced. RSSI was also integrated into other wireless data transmission protocols such as Bluetooth or ZigBee. The RSSI value does not represent directly the signal power but is a relative measure related to a reference value. Typically, the reference value $P_{ref} = 1$ mW is assumed, and RSSI is determined from the formula:

$$\text{RSSI} = 10\log_{10}\left(P_{RX} / P_{ref}\right) \text{ dBm} \qquad (4.2)$$

From Equations (4.1) and (3.2) it follows that the signal power decreases logarithmically with increasing distance.

The triangulation method enables evaluation of the absolute position of an object (a worker in this case) in a predefined coordinate system based on measurements of distance of the object from at least three signal transmitters deployed in the supervised area (Figure 4.3). Given coordinates x and y corresponding to positions of transmitters in space and the distances of the object to be located from each of the transmitters (determined from RSSI value measurement) it is possible, with the use of appropriate geometric relationships, to determine the coordinates of the position of the object. The accuracy to which the position of a worker can be determined will obviously depend on the precision with which the distance to individual transmitters can be evaluated. In some cases, such evaluations may be biased with a significant error due to currently prevailing radio wave propagation conditions. In real-life conditions, and especially in rooms, radio wave propagation is affected by a number of factors such as: reflections of the radio wave from metallic objects, interference with other waves, attenuation in mediums other than air, or refraction. Example methods aimed at improvement of the precision with which the distance is evaluated, consisting in taking into account certain environmental indicators concerning radio wave

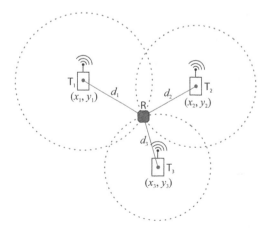

FIGURE 4.3 Determination of object position with the use of the triangulation method (O—the located object; (x_n, y_n)—transmitter position coordinates; d_n—transmitter-receiver distance).

propagation or applying adaptation optimization methods, are presented in the literature [Yang 2014; Kritz et al. 2016; Ma et al. 2017].

Detection of an event consisting in a worker entering a hazard zone is based on comparison of the worker's position coordinates with coordinates defining boundaries of the hazard zones. This requires obviously knowledge of the location of individual hazard zones, in particular establishing boundaries of the zones in the adopted coordinate system. The triangulation method enables evaluation of the worker's position relative to hazard zones which in general may represent figures of any geometrical shape. However, the method requires firstly, deployment of transmitters throughout the workplace area the number and parameters of which will be decisive for location accuracy, and secondly, more intense usage of network resources in view of the necessity to store data on position of transmitters and hazard zones and calculate position coordinates of workers moving in real time. An example schematic diagram of a network of that type is shown in Figure 4.4. The radio transmitters used for the purpose of location are Bluetooth beacons. Like in the previous example, the network works based on Wi-Fi standard and the worker whose location is to be established wears a wearable device being a network element.

The workplace area in which the considered sensor network operates is inscribed into an arbitrarily defined system of coordinates. The measuring devices (marked M in Figure 4.4) and RFID tags (R) included on the network are distributed at points coordinates of which in the adopted coordinate system are stored in a database of the network main unit. In the same database, coordinates of the points defining

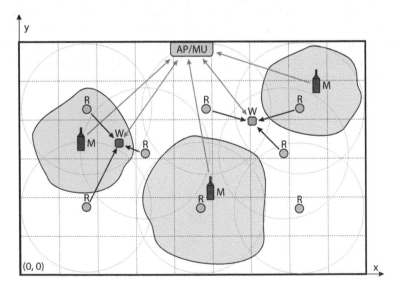

FIGURE 4.4 Structure and the principle of operation of a sensor network with determination of worker's location (M—measuring device/sensor; W—wearable device, R—RFID device/beacon; AP/MU—access point connected with main unit; black arrows—Bluetooth LE data transmission direction; gray arrows—Wi-Fi data transmission direction; x, y—coordinate system axes; gray areas—hazard zones; black dashed lines—coordinate system grid; gray circles—RFID range).

boundaries of hazard zones are also collected. A wearable device functioning in the network receives data concerning signal power (RSSI) from nearest RFID tags and conveys the data to the network main unit. The main unit, based on the received data (containing tag identification numbers and RSSI values) and the data on location of tags, determines position of wearable device (worker) in the coordinate system.

In up-to-date digitally controlled receiving modules, the RSSI value can be read out by using suitable software commands, making the data available to the user's application for the specific purpose the realized network is intended to serve. This fact is more and more universally made use of by various technical solutions to evaluate distance between objects or determine position of individual objects. A method of object localization based on RSSI values for signals received from Bluetooth LE (BLE) beacons was presented in Ma et al. [2017]. In an experiment carried out in a library hall with dimensions 18 m × 30 m, the average location accuracy of 0.87 m was obtained.

One method of improving the object position evaluation accuracy is the finger-prints technique [Yang 2014; Ma et al. 2017]. The method consists in using a plurality of transmitters deployed throughout the area in which the position of an object is to be traced. To implement the method, it is necessary to collect a database of model patterns describing the electromagnetic field in the object tracing area. These patterns serve as a kind of the electromagnetic field "fingerprint" of the space in question. The database of patterns is obtained by establishing a grid of points in the tracing area followed by determining RSSI values for each of the points from individual transmit-ters (Phase I of the fingerprint method application—the data acquisition). This way, a standard of the RSSI vs. position coordinates relationship is obtained. In the next step of application of the method (Phase II—object tracing) RSSI values read out by the traced object are compared with the values stored in the patterns database in search of values closest to those measured. As the RSSI values in the database are assigned to position coordinates, the actual position of the object can be determined. In one example of the application of this method to identify object positioning in buildings described in Kritz et al. [2016], Wi-Fi access points and Bluetooth LE beacons were used as transmitters. The receivers were mobile devices of various types (smart-phones). This method is time-consuming, especially in Phase I, and requires heavy use of network resources. However, the effort and cost results in a high level of object location accuracy, as the collected database of patterns takes into account the actual propagation pattern of radio waves in the tracking area.

4.3 THE USE OF RENEWABLE ENERGY SOURCES IN WIRELESS SENSOR NETWORKS

4.3.1 POWER DEMAND IN WSNs AND LIMITATIONS FOR THE USE OF RENEWABLE ENERGY SOURCES

One vital problem connected with design and construction of wireless sensor net-works pertains to supplying power for elements making up the network. A wireless sensor network is a set of many autonomous electronic devices energy consumption of which depends on their specific tasks, the way such tasks are accomplished, and the size of the area over which these devices are deployed. Provision of an adequate power supply for these devices is a key issue impacting proper functioning of the sensor network. The way this supply is provided has a direct effect on costs incurred for deployment and subsequent operation of the network.

The simplest solution would be to power sensor network devices directly from the electric power grid. Such solution minimizes the cost connected with sensor network operating maintenance at less structural and functional limitations resulting from the necessity to minimize the electric energy consumption by the devices—it is possible, for instance, to adopt less restrictive assumptions concerning the range and/or data transfer rate in radio transmission which have an important effect on electric energy demand. An obvious drawback of the solution, however, is the necessity of installing the devices at points with access to power grid or expansion of the grid suitably. Obviously, such a power supply method cannot be used for portable devices.

Another option which can be and actually is employed most commonly to power sensors in wireless networks are chemical cells or batteries. This type of power supply offers full independence and flexibility as far as the spatial distribution of sensor network devices is concerned and is always considered the first choice for any portable device. However, the use of a cell-based power supply involves the necessity to pay special attention to minimization of electric energy consumption in the individual devices. This also has an effect on the duration of maintenance-free operation of the network and on battery replacement or recharging costs (both economical and environmental).

The third option consists in using renewable energy sources (RESs) to power specific sensor network component devices [Morzyński 2015]. For acquisition of small quantities of electric energy from RESs to power autonomous measuring devices, the term "energy harvesting" was recently coined. In industrial conditions, three kinds of renewable energy are of practical interest: solar (optical radiation) energy, thermal (heat) energy, and mechanical (vibrational) energy. In the case of utilization of power sources of these types, the concerns are: the availability of the selected RES, its energy output, and its output variability over time. Electric power generators converting a given energy type into electric energy are characterized with parameters of the generated currents and voltages characteristic for the given generator type. The values and variability of the generated voltages differ significantly from those required to supply circuits of electronic sensors which are powered from direct current sources with the voltage falling, in most cases, in the range from 1.8 to 5 V (typically 3 V). Therefore, in electric power-supply systems which obtain energy from renewable sources it is necessary to use specialized electric energy-converting electronic circuits to secure adequate voltage parameters. Such converting circuits have a built-in function called maximum power point tracking (MPPT), as the maximum electric power supplied by a generator (e.g., a photovoltaic cell) depends on the load connected to it. In other words, a generator operates at maximum power when the current drawn from it has a specific value defined for given operating conditions.

Power sources for sensor circuits must meet their demands in the scope of the supplied current. For the majority of their operating time, sensor modules draw electric current in the range from a fraction of mA to several mA, however the current can increase up to several dozen mA at moments of transmitting radio messages. Small electric power generators which can be employed in sensor networks offer electric powers in the range from fractions to several dozen milliwatts. The current output of such generators is therefore insufficient to power wireless communication units directly. A solution to the problem is to store electric energy in periods when the sensor module shows lower demand for power and use the stored energy in periods of higher power consumption. The need to accumulate energy in power supply systems that use RESs follows also from the specificity of these sources which, although renewable, are by their nature not available continuously (e.g., solar energy is available only during the daytime). The electric energy in power systems may be stored in supercapacitors or in high-capacity rechargeable (e.g., lithium-polymer, Li-Po) batteries. Taking into account the fact that the majority of specialized electric energy converting circuits developed specifically for energy harvesting applications have a module for managing lithium-polymer cells already integrated in their structure, Li-Po batteries should be considered the preferred electric energy storage elements.

The following subsection contains a short characterization of electric power generators making use of the above-mentioned RES types, with stress put on their possible application in wireless sensor networks [Morzyński 2015].

4.3.2 Optical Radiation Energy Harvesting—Photovoltaic Cells

An electric power generator making use of optical radiation energy is the photovoltaic cell (known popularly as the solar cell or solar battery). Energy conversion in a photovoltaic cell occurs as a result of the photovoltaic effect (electromotive force in solid-state matter exposed to optical radiation). The market offers a variety photovoltaic cells differing in the employed technology, manufacturing cost, and effectiveness of the energy conversion process (in universally used cells, the conversion efficiency approaches 20%–25%). The electric power generated by photovoltaic cells is a product of conversion of electromagnetic radiation from a specific wavelength range depending mainly on the so-called *energy gap* characterizing the semiconductor from which the cells are made. For instance, monocrystalline silicon cells convert electromagnetic radiation with wavelengths from about 350 to 1200 nm, with the best efficiency observed for radiation with wavelengths of about 1000 nm. Thin-layered cells made of amorphous silicon convert electromagnetic radiation with wavelengths from about 300 to 800 nm, with the highest efficiency observed for waves with lengths of about 500 nm. As the popular name of the cells correctly suggests, they are in principle optimized with regard to conversion of solar energy contained in optical radiation—a significant portion of which falls not only in the visible wavelengths range but also in the infrared range. Obviously, photovoltaic cells are capable also of converting optical radiation energy from other sources of light but the power supplied will depend on the spectral characteristics of radiation emitted by such sources of artificial light.

The market offers a wide range of photovoltaic panels comprising one or more photovoltaic cells with a variety of dimensions. In general it can be assumed that the electric power generated by a panel increases with its surface area. It is therefore possible to find a photovoltaic panel capable of generating power sufficient to supply given sensor module at known illumination conditions. On the other hand, the use of a panel dimensions of which would increase significantly the size of the combined device would restrict its functionality to a degree unacceptable in most cases.

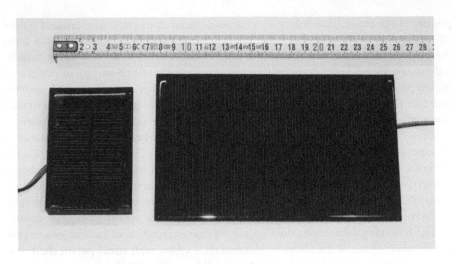

FIGURE 4.5 Photovoltaic panels 1 (left) and 2 (right).

Figure 4.5 show examples of small photovoltaic panels with dimensions 92 × 61 mm (panel 1) and 180 × 112 mm (panel 2), which could be used to power sensors in wireless sensor networks.

Table 4.2 summarizes results of electric power measurements taken for the two photovoltaic panels under conditions of two purely resistive loads of 200 and 1000 Ω. The panels were operated in the following illumination conditions:

- illumination with solar light penetrating into a room through a window at noon, on a cloudy day, at the illuminance of 1050 lx,
- illumination with light of fluorescent lamps installed in a laboratory room, at the illuminance of 680 lx, and
- illumination with incandescent light emitted by a laboratory lamp with halogen bulb at the illuminance values of 200 lx, 2000 lx, 4000 lx, and 8000 lx.

TABLE 4.2

Electric Power (in mW) Supplied by Photovoltaic Panels 1 and 2 Depending on Load and Illumination Conditions

	Panel 1		Panel 2	
Illuminance	Load 200 Ω	Load 1000 Ω	Load 200 Ω	Load 1000 Ω
Solar light 1050 lx	0.100	0.353	1.114	4.077
Fluorescent lighting 680 lx	0.014	0.050	0.160	0.591
Halogen lighting 200 lx	0.316	1.134	2.736	10.985
Halogen lighting 2000 lx	12.469	15.66	85.9314	61.082
Halogen lighting 4000 lx	46.190	19.158	171.60	67.396
Halogen lighting 8000 lx	91.938	21.534	278.082	68.657

For comparison it is worth keeping in mind that the illuminance values recommended for workstations are: 200 lx for workshop activities, 500 lx for office work, and 1000 lx for precision work.

The presented results of tests performed with the two photovoltaic panels indicate that in the case of illumination with fluorescent lamps (typical for industrial shop floors), the obtained electric power values were less than 1 mW. Significantly higher powers can be obtained with natural or incandescent light. Even illumination with diffused sunlight on a cloudy day, the obtained electric power values exceeded 4 mW for panel No. 2. Using more intense incandescent lighting with illuminance of 2000 lx, electric power values increased to above 85 mW (panel No. 2) and 15 mW (panel No. 1). Under very intense incandescent lighting with the intensity of 8000 lx (Table 4.2), the power generated by the larger panel was 278 mW compared to 92 mW generated by the smaller panel. The presented results indicate that photovoltaic panels are best suited for supplying power to sensor modules used outdoors or in rooms with either good sun exposure or incandescent lighting.

4.3.3 MECHANICAL VIBRATION ENERGY HARVESTING—INDUCTION AND PIEZOELECTRIC GENERATORS

Vibrating objects or components of machines and devices which are normally regarded as sources of vibro-acoustic hazards in the work environment, can be also used as a source of energy to power sensor modules. To convert the energy of mechanical vibrations into electric energy, induction generators or piezoelectric generators may be utilized [Morzyński 2015].

In induction power generators, the electromagnetic induction effect is due to the development of electromotive force in a conductor as a result of variations in the magnetic induction flux encompassing the conductor. Such magnetic flux induction value change can be caused by motion of the conductor and the magnetic field source relative to each other. The value of the induced electromotive force (voltage value) depends on the magnetic induction flux variation rate.

Example induction generator design solutions are shown in Figure 4.6. Structure of the device comprises typically an induction coil wound onto a tube inside which, as a result of vibrations, a neodymium magnet is moving. The magnet may be suspended on a spring or in a magnetic field of another immobile magnet. The maximum electric power values supplied by the induction generators shown in Figure 4.6 fall into the range of 1.3–2 mW.

Piezoelectric generators are devices in which piezoelectric materials or piezoelectric transducers are used. The piezoelectric effect occurring in such materials or structures consists in the appearance of uncompensated electric charge on their surfaces as a result of mechanical stresses induced by external forces. A piezoelectric generator will be therefore capable of using mechanical vibrations to induce stresses in the piezoelectric transducer installed in it as a result of which it will become a source of power supply voltage. The electric voltage generated by the presented transducer is proportional to amplitude of stresses induced in its crystalline material. The transducer must be therefore fixed (glued)

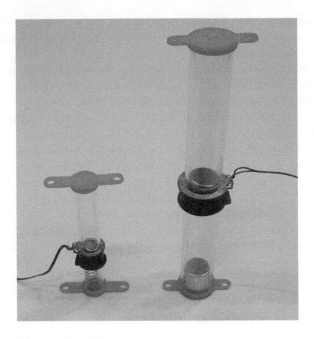

FIGURE 4.6   Induction generators.

to a mechanical structure vibrations of which will cause deformations resulting in occurrence of stresses in piezoelectric material.

Example piezoelectric generators constructed of composite piezoelectric transducers glued onto a cantilever beam are shown in Figure 4.7. Electric power values of the generators approached 0.15 mW.

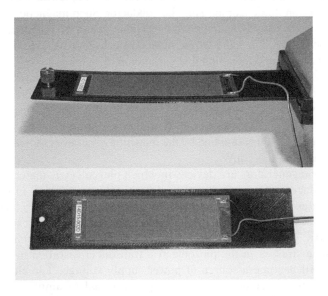

FIGURE 4.7   Piezoelectric generators.

A characteristic feature of generators converting the energy of mechanical vibrations into electric power is the alternated nature of the output voltage. The maximum power output of such generators is observed for the excitation frequency equal to the eigenfrequency of the device. That means that the generator, through proper choice of its mechanical parameters, must be tuned to the source vibration frequency. Otherwise, the available electric power dramatically decreases. Numerous studies are continued worldwide on generators with nonlinear characteristic which would be capable of converting effectively mechanical energy into electric energy in wider vibration frequency bands (cf. e.g., [Wang et al. 2017]). Example applications of electric power generators in which energy of mechanical vibration is used are given in [Dziadak et al. 2016].

4.3.4 THERMAL ENERGY HARVESTING—THERMOELECTRIC GENERATORS

Thermal energy, a measure of which is the temperature of an object, can be generated as a byproduct of industrial processes or operation of individual machines and devices. The source of the energy can be any combustion process (burning fuels in internal combustion engines), thermal treatment of materials, compression of process gases, or electric appliances such as electric motors. Only in rare cases is such energy reclaimed and redirected for other purposes (e.g., heating rooms). In the majority of cases, thermal energy (in view of its undesired effect on the course of technological processes or operation of machines and absence of economically viable recovery methods) was lost or eliminated by using suitable cooling systems. However, with currently available technologies, thermal energy can be used to produce electric power to a much larger extent [Morzyński 2015].

An example of an electric generator capable of converting thermal energy into electric energy are Peltier cells offered on the market in the form of Peltier modules. Examples of commercially available Peltier modules are shown in Figure 4.8.

The Peltier junction is a structure (used most commonly in cooling devices) in which the so-called Peltier effect is observed as absorption of thermal energy when a voltage is applied to and electric current flows through a junction formed between two different semiconductor materials. A reverse physical phenomenon known as the Seebeck effect may also occur in the same device when electromotive force is developed in a system composed of two semiconductors, temperatures of which differ at the junctions. The market, apart from standard Peltier devices used in the refrigeration engineering, also offers Peltier modules manufactured specifically for applications in electric power generators.

The voltage generated by a cell composed of two semiconductors A and B is given by the formula:

$$V = \left(S_\mathrm{A} - S_\mathrm{B}\right) \cdot \left(T_2 - T_1\right) \, \mathrm{V} \tag{4.3}$$

where S_B and S_A (in V/K) are Seebeck coefficients characterizing two semiconductors making up the cell, whereas T_2 and T_1 are temperatures at the junction of

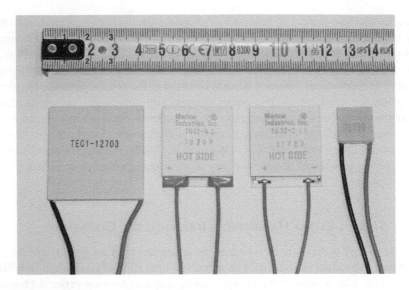

FIGURE 4.8 Examples of Peltier modules.

the semiconductors (in case of Peltier modules these are temperatures of opposite module surfaces). By supplying thermal energy to (heating up) one side of a Peltier module and dissipating the thermal energy transmitted to (cooling down) the other side one obtains a thermoelectric generator which can be used as a power source to supply measuring circuits.

The electric power output of a thermoelectric generator designed on the Peltier module will depend on the difference between temperatures of its "hot" side and "cold" side. For that reason, the "hot" side of the Peltier module should be equipped with an element improving effectiveness of heat exchange with the thermal energy source, whereas the "cold" side should have a system improving dissipation of energy into the environment (i.e., an effective radiator). Examples of thermoelectric generators constructed according to that principle are shown in Figure 4.9 [Morzyński and Szczepański 2017; Szczepański 2017; Szczepański and Morzyński 2017].

The generators were constructed with the use of type TG12-4L (Marlow Industries) Peltier modules optimized with regard to production of electric power. The modules are designed for applications in which the hot side temperature does not exceed 200°C in case of continuous operation, with 230°C being the short-term (peak) temperature limit. One of the generators is provided, on its hot side, with a heat-conducting element the shape of which enables attachment of the generator to any flat surface with magnetic properties (magnetic mounting with the use of neodymium magnets), while the other is designed to be mounted on pipe-shaped elements. To carry the heat from the cool side away, computer radiators with heat pipes were used. To ensure sufficiently effective thermal insulation between the hot side and the cold side, the cell was wrapped with ceramic fiber paper and Teflon plate around its perimeter. In order to improve the heat exchange, a heat-conducting

FIGURE 4.9 Thermoelectric generators designed to be mounted on flat surfaces (left) and on pipes (right).

paste was applied between the Peltier module, the radiator, and the base drawing heat energy from the source. The structures were integrated with the use of screw joints, with Teflon washers placed under screws to minimize the quantity of heat supplied to the radiator via the screw joint.

Example results of tests performed with a generator intended for mounting on flat surfaces are presented in Figure 4.10.

The tests were carried out in a closed room at ambient temperature of 28°C. A heating table with controlled temperature was used as a heat source. The power capacity of the generator was determined for resistance loads in the range from 1 to 1000 Ω. The temperature differences between the cold and the hot side of the generator obtained in steady-state conditions are presented in Figure 4.10.

The highest power values are obtained when the generators are loaded with low-value resistance receivers and voltages generated at their outputs are included in the range from 0.2 to 1.2 V. When applied in power supply systems, the generators need to be equipped with electric power conversion circuits with MPPT capacity and a voltage booster. Even for small temperature differences between the hot side and the cold side of the generator, it is possible to obtain electric power values up to 10–20 mW which is sufficient to supply some sensor modules. The power values exceeding 90 mW can be achieved at higher temperature differences and are adequate to power sensor modules in which the electric power demand is relatively high. From the above-described merits, it follows that thermoelectric generators have the potential to be widely used as power sources for certain immovable components of wireless sensor networks in industrial conditions. Figure 4.11 presents an example design solution of a noise sensor powered by a thermoelectric generator.

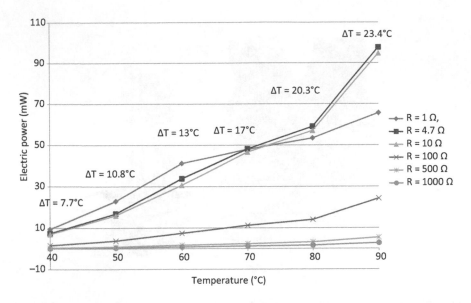

FIGURE 4.10   Results of examination of an electric thermoelectric generator.

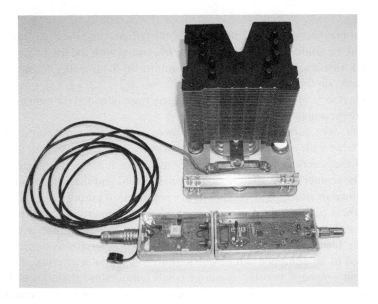

FIGURE 4.11   A wireless noise sensor powered from a thermoelectric generator.

The presented device constitutes an element of a sensor network used to measure noise and vibration levels in the work environment [Morzyński and Szczepański 2017] with wireless communication based on ZigBee standard. The power system of the sensor module was constructed based on bq25570 integrated circuit (nano power boost charger and buck converter for energy harvester powered applications) marketed by Texas Instruments.

4.4 EXAMPLES OF MODEL SENSOR NETWORK SOLUTIONS IN NOISE HAZARD REDUCTION

4.4.1 A SYSTEM FOR REMOTE SUPERVISION OVER CORRECT USAGE OF EARMUFFS

Personal hearing protection devices—such as earplugs and earmuffs—are typically used when it is impossible to reduce exposure of individual workers to noise with the use of other technical means and organizational measures. The employer is obliged to ensure that workers use the personal protection equipment correctly. Hearing protection devices should be selected with parameters of the noise existing at the workstation taken into account in a way guaranteeing that individual exposure to noise does not exceed the permissible value. However, studies carried out in actual industrial site conditions [Berger 1998; Burks and Stein 1998; Murphy and Franks 2000; Kotarbińska et al. 2007; Canetto 2009; Kotarbińska and Kozłowski 2009] indicate that the actual sound attenuation provided by earmuffs in real-life conditions can be a dozen or so decibels lower than expected from theoretical calculations based on results of noise measurements and noise reduction values determined as part of equipment certification tests. In extreme cases, individual exposure to noise may exceed the permissible exposure values. The underlying cause of such divergences can be incorrect usage of earmuffs. In practice, workers are not always familiarized sufficiently well with rules and correct methods of using earmuffs or sometimes ignore them consciously (failing to take into account possible consequences for their health). High noise-related hazard levels in the work environment and discrepancies between the predicted and the actual noise level under ear cups of earmuffs (for reasons explained above) may have serious consequences to workers' health.

One of the possible solutions to this problem may be the development of a wireless sensor network to supervise the usage of earmuffs through continuous monitoring of noise parameters under earmuff cups [Morzyński 2013, 2014]. Today's market of hearing protection devices offers a wide range of earmuffs with electronics such as: level-dependent earmuffs or earmuffs with active noise control and/or radio communication. It is therefore possible to design earmuffs with noise measuring circuits (noise sensors) which might be configured as elements of a wireless sensor network. Figure 4.12 shows a schematic diagram of one basic structure proposed to enable remote supervision over correct usage of earmuffs. The system comprises a plurality of earmuffs with integrated noise measuring devices (sensors) forming a sensor network which communicate with a main unit using one of wireless data transmission standards.

Operation of the system consists in continuous monitoring of noise levels under ear cups of earmuffs integrated in the system, recording parameters of noise under earmuff cups, and alerting both the system supervisor and the worker wearing the device to occurrence of excessive noise exposure and the need for suitable corrective action. The role of measuring devices installed in earmuffs is to continuously measure parameters of noise under earmuff cups, determine the values of physical quantities characterizing exposure to noise in the working environment, and compare the values with permissible values. Data on noise collected from measuring

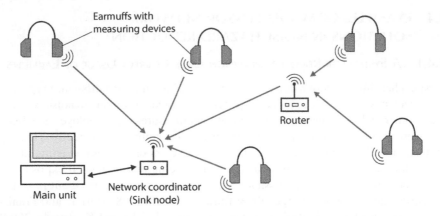

FIGURE 4.12 A schematic diagram of a system for remote supervision over correct usage of earmuffs.

devices installed in earmuff cups are conveyed via the wireless network to the main unit of the system. The function of the main unit includes processing the incoming data on noise parameters with regard to existing or potential hazards and advising the system supervisor of the necessity of appropriate action. If necessary, routers can be also used in large-area wireless networks. A system constructed this way enables management to keep permanent control of correct use of anti-noise earmuffs and allows supervisors to take prompt corrective action in case of irregularities and excessive noise exposure for individual workers.

The system in question should continuously evaluate noise quantities important for hearing protection. These quantities are:

- $L_{EX,8h}$—the A-weighted noise exposure level normalized to an 8 h working day (daily noise exposure level), and
- $L_{p,C,peak}$—the C-weighted peak sound pressure level.

Exposure of a worker to noise must not exceed the permissible values set out for these quantities (the values may be different, depending on the country of application) [Directive 2003/10/EC, NIOSH 1998]. For that reason, the system should alert to any instance of any of the permissible values being exceeded.

The system should monitor values of the above-listed quantities and/or detect the fact that the permissible values established for any of the quantities was exceeded, by processing the measured acoustic signal and performing relevant numerical calculations throughout the exposure period. The value of the A-weighted noise exposure level normalized to an 8 h working day is usually determined based on A-weighted sound pressure level or the equivalent continuous A-weighted sound pressure level measurements which involves execution of a number of complex arithmetic operations. For the proposed system it is more convenient to use an equivalent quantity, the *sound exposure E_A* [Berger et al. 2003]. It can be calculated based on short-time measurements of the root mean square value of A-weighted sound pressure p_A. The sound exposure can be calculated in the measuring module and sent to the main unit after defined measurement intervals or it can be calculated synchronously in the main unit based on transmitted p_A values.

The value of C-weighted peak sound pressure level can be determined based on C-weighted peak sound pressure measurements.

In view of the worker protection purpose of the system and the possibility of one of the earmuff cups being damaged or incorrectly worn (resulting therefore in overexposure of one of the worker's ears), it is necessary to carry out measurements and related calculations for each of the earmuff cups independently.

Measuring devices designed to be installed in earmuff cups should be battery-powered. For that reason, the measuring devices should meet the following requirements:

- both dimensions and the mass of the device should be reduced to a minimum,
- the device should operate incessantly for at least 8 h without battery replacement or recharging, and
- the structural design of the device should be optimized with regard to energy consumption to maximize the operating time of the device and minimize battery size and weight.

To measure the quantities characterizing noise in the work environment correctly, the measured acoustic signals must be filtered with the use of A- or C-weighing filters. Filtration of the measured sound signals can be effected by means of digital processing of the signals or with the use of analog filters. Digital filtration would simplify significantly the structure of the measuring module, however it would require the use of a microcontroller with sufficiently large computing capacity which would increase the demand for electric power. For that reason, the analog method for processing the measurement signal was used in the described system.

When making the choice as far as the wireless transmission standard for the system is concerned, the following assumptions were taken into account:

1. The workplace management's intent on deploying the remote supervision system may differ significantly in terms of the number of employees, structure, and size (in the meaning of the surface area of the work establishment site). For this reason, the wireless transmission standard chosen should enable the creation of both small simply-structured wireless networks and vast multi-level networks encompassing extensive areas and large number of measuring modules. Significant features will therefore include the maximum number of nodes which can be served in given transmission standard (from several to several hundred) and the maximum distance at which the data transmission can be effected with possible obstacles in the form of walls and other structural elements of individual workrooms taken into account

(the signal propagation range should be adequate to ensure undisturbed transmission of data within the area of a typical industrial shop floor).
2. Measuring devices comprising a data transmission module in their structures will be integrated into earmuffs and will be battery-powered. For that reason, they should be as small and light and consume as little power as possible, so that the measuring device can be integrated into an earmuff cup and run continuously for at least 8 h. The data transfer rate characterizing the network is not a critical parameter and the minimum requirement does not exceed 50 kb/s, therefore the transmission standard should be chosen and the radio devices designed with electric energy saving procedures taken into account.

In the final analysis of this situation, ZigBee was chosen as the standard most suitable for this application. From appropriate model options (compatible with ZigBee standard) offered on the market, XBee series 2 (Digi International) radio circuits (modules) operating in the ISM 2.4 GHz frequency band were chosen to realize the presented system. Depending on the software employed, each of the modules can function as a coordinator, a router, or a ZigBee end device. A wide variety of XBee modules is offered differing in programming capabilities, output power, and antenna solutions. The following solutions were applied in the structures developed for the purpose of the network:

- XBee module with a wire antenna to construct measuring devices in earmuffs,
- XBee Pro module with RP-SMA antenna connector to construct routers, and
- XBee Pro module with UF-L antenna connector to construct the network coordinator functioning also as the sink node connected to the system main unit.

All the modules utilized are Series 2 devices.

A block diagram of the measuring module is presented in Figure 4.13. The module comprises power supply circuits, two independent analog measuring channels for each of earmuff cups, and the measurement data processing and transmission circuit in the form of XBee module. Signals from outputs of analog circuits are connected directly to the inputs of A/D converters of the XBee module. In the development of the measuring device, a simplifying assumption was adopted—consisting of taking

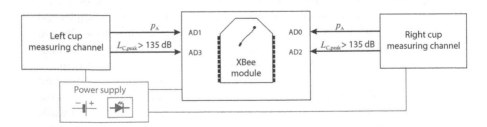

FIGURE 4.13   A block diagram of measuring device with XBee module.

measurements of the C-weighted peak sound pressure level. The measuring device does not determine the value of the parameter but rather checks whether the permissible value is or is not exceeded. Such information is sufficient to assess correctness of usage of anti-noise earmuffs and at the same time allows significant simplification of the system structure.

Outputs of the measuring channels correspond to rms A-weighted sound pressure values and signals alert to events where the permissible value of the C-weighted peak sound pressure level was exceeded. The signals are supplied to inputs of analog-to-digital converters of XBee module according to the above-presented schematic diagram. The XBee module, at predetermined intervals (e.g., every one second), samples and converts signals reaching the converters and then transmits them to the main unit of the system.

The measuring device (comprising a XBee module) was actually built into two printed-circuit boards (PCBs) given shapes adapted to earmuff cups. The left-hand-side board (Figure 4.14) contains the measuring channel for the left earmuff cup together with elements of the power supply system corresponding to that channel. The right-hand-side board is somewhat larger and comprises the measuring channel for the right earmuff cup and the XBee module together with elements of their power supply circuits. This PCB structure was designed to minimize interference caused by the Xbee module in the measuring channels.

A view of the model earmuffs with measuring devices of the system mounted is shown in Figure 4.15.

The coordinator and routers of the wireless data transmission network were constructed based on properly programmed XBee Pro Series 2 modules. The modules were completed with necessary peripheral circuits enabling them to function as autonomous devices and mounted in suitable housings (Figure 4.16).

According to the adopted system architecture, the main unit of the system is a PC-class computer with specifically developed software functions for use as a device collecting and processing data on noise exposure in individual workers wearing earmuffs. To a USB connector provided in the main unit, a ZigBee network coordinator is connected functioning as a sink node via which data from measuring devices in

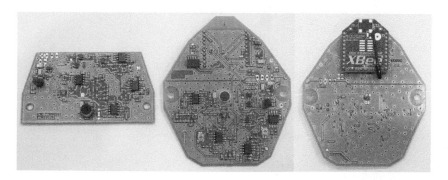

FIGURE 4.14   A view of printed-circuit boards of the measuring device with XBee module, showing from left to right: the l.h.s. board (top view), the r.h.s. board (top view), and the r.h.s. board (bottom view).

FIGURE 4.15 The model earmuffs with integrated XBee module-based measuring device.

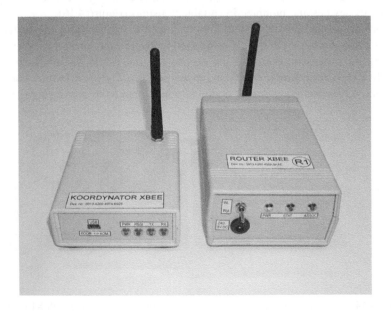

FIGURE 4.16 ZigBee network model coordinator and router constructed with the use of XBee modules.

the hearing protection aids are transmitted to the main unit. The software developed for the main unit performs the following functions:

- collects data on noise level from the ZigBee network coordinator and stores this data in a database,
- analyzes (in real-time mode) the received data points by comparing them with permissible values and generates warnings of hazard occurrences for the system supervisor, and
- processes and presents (in tabular and/or graphical form) the collected data upon request of the system administrator.

The software developed to manage the remote supervision system is written in Java language and exchanges data with MySQL database software. The software enables information concerning earmuffs and their users to be entered into the database and previewed directly (in real time). The measurements for a selected earmuff including values of the A-weighted sound pressure level and the sound exposure (together with information about possible instances exceeding the C-weighted peak sound pressure level permissible value) are reported in the form of tables and/or histograms. The software also offers the option to browse through and analyze information concerning the exposure to noise determined with the use of measurement results stored in the system database. Figure 4.17 shows a view of the program home window with a tab showing results of noise measurements currently realized in the network by measuring devices installed in earmuffs. Another tab titled "Wyniki" (Results) offers browsing through noise measurement results stored in the system database. The "Baza" (Base) tab is used to maintain the database of workers and hearing protectors supervised by the system. A menu available at the tab top contains software commands concerning calibration of measuring devices.

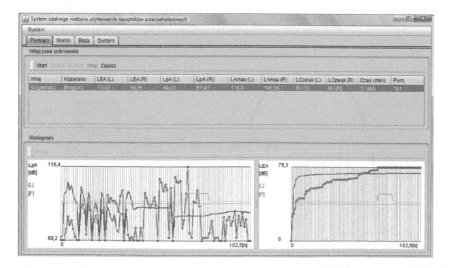

FIGURE 4.17 The program home window with the "Pomiary" (Measurements) tab opened.

This model of the system was subjected to tests under real-life conditions prevailing in a forging shop of a large industrial workplace. The noise sources present on the shop floor were the production equipment items such as die forging hammers, presses, rolling mills, and heating stoves equipped with vibration feeders for heat-treated articles. The other noise sources included numerous fans cooling the shop floor and operator's workstations, the processed steel articles in the course of handling (tipping over and tossing into boxes for finished parts, etc.), and forklift trucks moving around the shop floor. In the course of tests, measurement results acquired with the use of the system were compared with results obtained with the use of SVAN 948 class 1 sound level and analyzer (Svantek). A schematic diagram of the measuring setup is presented in Figure 4.18. A miniature microphone used typically in measurements carried out with the use of MIRE (Microphone In Real Ear) technique was mounted in the earmuff cup close to the measuring device microphone and connected to the reference sound level by means of a cable. During the tests, variations of the A-weighted sound pressure level (over the course of measurements) and the C-weighted peak sound pressure level (or information of the permissible value of the latter being exceeded) were recorded both in the system and in the reference meter.

Results of test measurements from this example are presented in Figure 4.19 in which the upper panel represents results observed in the tested system and the lower panel shows the corresponding data recorded with the use of SVAN 948 reference meter. The upper graph shows A-weighted sound pressure level waveforms for the right-hand-side (dark gray line) and the left-hand-side (light gray line) earmuff cup. The SVAN 948 meter was used to take measurements in the right-hand-side cup of the prototype earmuffs. Graphs on lower panels present data recorded with the use of SVAN 948 meter, namely the course of the A-weighted sound pressure level waveform (light gray line) and the C-weighted peak sound pressure level (dark gray line). Graphs shown in the upper and lower panel are not synchronized in time because of different moments at which the measuring devices switched on; however, the data can be used for the purpose of comparison of the nature of the A-weighted sound pressure level variability in time. To compare operation of

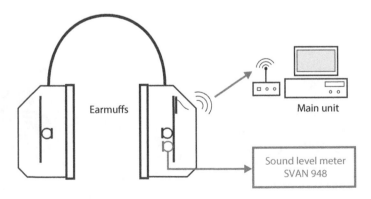

FIGURE 4.18   A schematic diagram of the measuring setup in the course of measuring device tests.

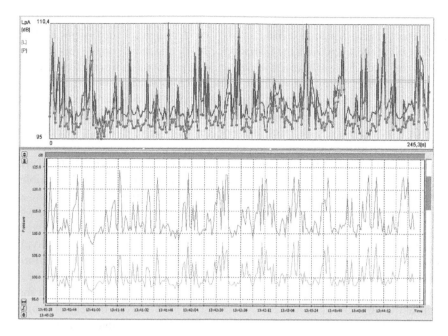

FIGURE 4.19 Example results of measurements taken with the use of the developed system (upper panel) and SVAN 948 meter (lower panel).

measuring devices installed in earmuffs with operation of the reference system for A-weighted sound pressure levels falling in the range of correct operation of the measuring devices installed in the earmuffs (in this case, from 75 to 115 dB as it follows from the processing dynamics range characterizing the used measuring devices), the tests were also carried out with earmuff cups unobstructed (i.e., the earmuffs put down or carried freely). That way, high sound pressure levels of the noise reaching the measuring microphone of the device were obtained. The plot of Figure 4.19 illustrates the transient nature of the noise due to the cyclic nature of operation of die-forming hammers and processed article feeders.

The results of these tests show that this system correctly measures and reproduces the quantities characterizing noise penetration under an earmuff cup. The wireless sensor network of this type can therefore be regarded as a solution to problems connected with correct usage of earmuffs—having a positive effect on improvement of safety among employees working in conditions involving noise hazards.

4.4.2 Wireless Sensor Network for Monitoring Work Environments and Alerting Workers to Hazards

Another example designed to support reduction of workers' exposure to hazardous factors existing in the work environment (including noise) is a sensor network designed as a tool for monitoring the work environment and alerting workers to specific hazards.

Applicable laws and regulations obligate employers to provide their employees with safe working conditions, including by measurement of factors posing health hazards in the work environment [Koradecka 2010] and reduction of workers' exposure to any such factors levels of which exceed permissible limits. Concentration or intensity of hazardous factors in the work environment may be subject to variations resulting from changes in work process parameters, normal or excessive wear of machines and tools, or unexpected emergency conditions resulting in increased hazards to workers. In such cases, the workers may be exposed to the effect of a harmful factor in the work environment with intensity or concentration exceeding the permissible values. Identification of such a hazard—as well as prompt and effective response aimed at reduction of its effect on worker's health—is made possible by continuous measurement and monitoring in the work environment. There may also be workstations at which a worker may migrate between different areas of the workplace being thus exposed to various hazards. Continuous monitoring of work environment parameters enables improved assessment of hazards to which individual workers are exposed and facilitates the development of actions aimed at minimization of the effects of these hazards. Such tasks can be accomplished by a sensor network. Moreover, continuous monitoring of work environment parameters (at many points at the same time and realized by means of a network) can be helpful in understanding and assessing processes occurring in the work environment and the related hazards to workers with regard to development of effective strategies and measures aimed at reduction of the hazards.

The proposed sensor network [Morzyński 2019a, Morzynski 2019b] comprises a plurality of autonomous measuring devices (sensors) enabling monitoring of hazards existing in the work environment and further, a number of wearable devices worn by workers and conveying to them information about hazards existing at the location they find themselves at the moment. The network structure is presented in Figure 4.20. The Wi-Fi standard was chosen for communication between network elements. This facilitates connecting the network to the Internet and thus, for instance, managing its operation from a remote computer.

In the proposed network the workers, thanks to the use of wearable devices and based on information conveyed to them via the network, can modify their behavior and adapt it to conditions prevailing in the work environment (by avoiding hazard zones or wearing suitable personal protection equipment). It is obvious that workers should be alerted to a hazard only in case of entering a zone in which the hazard actually exists. That means the necessity to locate workers within the workplace area and evaluate their position relative to hazard zones. The location method employed in the proposed solution is based on measuring the power of received radio signals and the triangulation method described earlier in Section 4.2. The transmitters used for location determination are Bluetooth LE beacons. To ensure versatility of a sensor network—allowing it to be used to monitor hazards of various types—a structural assumption was adopted whereby the measuring (sensor) elements of the sensor network have a modular structure. Connection between the sensor module and the communication module is provided by means of a suitable interface which joins them mechanically and connects electrically at the same time. In the following, details of example solutions are presented.

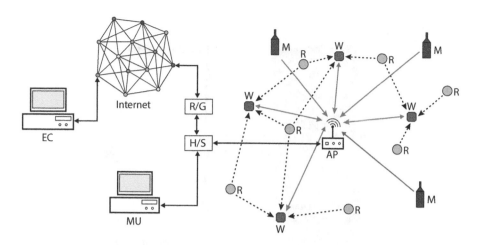

FIGURE 4.20 Basic structure of a sensor network used to monitor the work environment and alert workers to hazards (M—measuring device/sensor; W—wearable device; B—Bluetooth beacon; AP—Wi-Fi access point; MU—network main unit (computer); H/S—hub/switch; R/G—network router/gate; EC—external computer; Gray arrows—Wi-Fi wireless transmissions; black dashed arrows—Bluetooth LE wireless transmissions; black arrows—Ethernet connections).

The radio transmitters used for the purpose of object location are Bluetooth LE beacons. Construction of network transmitters is based on iNode devices offered by a domestic company (ELSAT s.c.). The iNode beacon is a small-size radio module with the diameter of 28 mm, the standard version of which is powered from a 3 Volt type CR2032 cell. The cell ensures prolonged operation of the radio unit at standard settings (i.e., the transmitted signal power of –2 dBm (33%)) and the transmission interval of 1.25 s (i.e., transmission session frequency of 0.8 Hz). From the point of view of accuracy and speed at which the position of a worker within the work establishment site is determined, it is favorable to keep the transmission interval as short as possible and (in case of the majority of actual network solutions) the transmitted signal power as high as possible which carries with it higher electric energy consumption. On that account, for the purpose of realization of the project, the present author's own structural solutions of radio transmitters were developed comprising an iNode beacon module with elaborated power units based on AA cells. A typical CR2032 has the capacity of 210 mAh whereas the capacity of AA cells reaches 2300 mAh. The employed solution enables the extension of the radio module operating time by up to a factor of ten. The power system comprises three cells connected in series and a linear voltage regulator of LDO type with the output voltage of 3 V. On the front side of beacon housing, a LED is provided signaling that the module is active. The housing is made using a 3D printer and is provided with neodymium magnets enabling the beacon to be fastened easily to any steel surface in the work environment. A view of the beacon is shown in Figure 4.21.

The tasks assigned to the wearable device include alerting the worker who wears it to existing hazards and providing data necessary to determine position of the

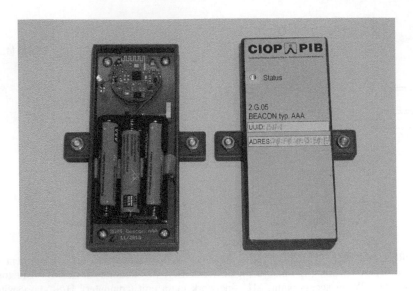

FIGURE 4.21   Bluetooth LE beacon with its housing closed and open.

worker relative to hazard zones. Such a wearable device should be a wrist-mounted apparatus, having the form of a wristwatch, with structural components comprising modules and elements for: hazard signaling by various means (for example, vibration exciter, LED, loudspeaker); providing information about the hazard in text form (display); user's response indicator (button); microcontroller, radio circuits for data transmission (Wi-Fi) and location determination (Bluetooth LE); and a power supply system. The form of a wristwatch and the use of a vibration exciter as one of hazard signaling elements were intentional as in noisy conditions, haptic signals seem to be the best perceptible ones.

Analyzing the market offerings of technical solutions which could be used to construct such a wearable device, attention was drawn to more and more popular dual-system radio modules (electronic devices capable of operating both in Wi-Fi standard and Bluetooth LE standard). Application of such solutions would enable significant simplification of the wearable device structure (one radio module instead of two). Finally, it was decided to base the wearable device structure on ESP32-WROOM-32 radio module, a popular module based on type ESP32 microcontroller SoC (System-on-a-Chip) circuit. Outside the module, ports of the microcontroller ESP32 are led out including the terminals of serial interfaces (HSPI, VSPI, I2C, UART), A/D converter inputs, D/A converter outputs, and general-purpose inputs/outputs (IO). This enabled the construction of a wearable device based on the same radio module without the need to use an additional microcontroller. To the radio module, a 0.95″ OLED-type display is connected on which messages for the wearable device user are displayed. In the power supply circuit, a Li-Po battery was used and an adjustable low voltage synchronous boost converter TPS61200 (Texas Instruments) enabled a stable power supply voltage of 3.3 V. The hazard signaling elements were connected to I/O ports of the radio module. The electronic circuit

FIGURE 4.22 A view of the wearable device electronic component: S—switch, U—USB mini connector; B—user's button; L—LED; O—OLED display; RM—ESP32-WROOM-32 radio module.

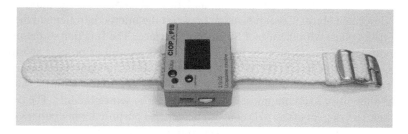

FIGURE 4.23 A wearable device—an overall view.

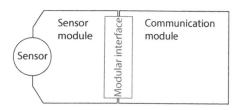

FIGURE 4.24 An overall block diagram of the measuring device.

board of the wearable device is presented in Figure 4.22, and its view is shown in Figure 4.23. The wearable device is programmed with the use of the Arduino programming environment.

In line with the adopted assumptions, the measuring device has a modular structure (Figure 4.24). It is comprises a sensor module with a sensor capable of measuring various hazardous factors, a communication module providing wireless data transmission within the network, and a power supply for the measuring device. Electrical coupling and mechanical joint between the modules is provided by a modular interface. The communication module, thanks to the use of the modular interface, can recognize automatically the measuring module connected to it and thus correctly receive and process data provided by the module. As a result, the measuring part of the sensor network may be configured and modified in a much simpler way. Such structure of the measuring devices enables also use of any communication module interchangeably with different measuring modules and thus facilitates and reduces the cost of possible repairs and maintenance of the sensor network.

The communication module functions include:

- providing power supply for the measuring device as a whole,
- identification of the connected sensor module,
- controlling the measurement process,
- measurement data processing and formatting for transmission, and
- transmitting measurement data via wireless Wi-Fi link to the sensor network main unit.

The main components of the communication module structure are microcontroller STM32L476VGT6 (in LQFP100 housing) and the ESP32-WROOM-32 radio module already described above. Communication between the microcontroller and the radio module takes place with the use of SPI serial interface. The function of the module microcontroller is also to identify the sensor module as well as process and format the data, whereas the task of the Wi-Fi radio module is to transmit the measurement data. The communication module microcontroller is connected with the modular interface allowing it to communicate with circuits of the sensor module. The connection includes analog inputs to A/D converters, general purpose inputs-outputs GPIO, SPI interface, and I^{2}C interface. As a power source for the module, a lithium-polymer (Li-Po) battery was used with the capacity of 3500 mAh and TPS61200 voltage boost converter (Texas Instruments). The battery is charged via USB mini connector and charging controller MCP73831. The housing was made with the use of the 3D printing technique. On its sides, it is provided with holders for neodymium magnets which are used to attach the module to any surface showing magnetic properties. A view of the communication module is presented in Figure 4.25.

The modular interface was constructed based on a printed-circuit board with two types of signal connectors and one type of connector providing the mechanical joint. Figure 4.26 shows a view of the modular interface.

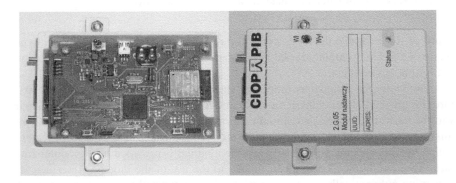

FIGURE 4.25   A view of the communication module with its housing open and closed.

FIGURE 4.26 The modular interface (the nearer portion) in the communication module (the further portion): P—interface printed-circuit board; H—Harting connector; B—banana connector; PHA and PHB—pin headers.

In the interface, a 20-pin type M55-7012042 Harting connectors (male, in the communication module) and M55-6012042 (female, in the measuring module) is mounted in central part of the module interface printed-circuit board. The connectors provide electric coupling of signals between the modules. An additional mechanical coupling of the modules is secured by banana-type connectors with the diameter of 2 mm mounted in the interface—female (sockets) in the communication module and male (plugs) in the measuring module.

The example noise sensor module presented in this chapter was constructed based on STM32L476RGT6 microcontroller and digital microphone MEMS type MP34DT05A (STMicroelectronics). The microphone is characterized by the following parameters:

- the maximum sound pressure level value (Acoustic Overload Pressure, AOL, the sound pressure level above which signal distortions exceed 10%)—122.5 dB,
- signal-to-noise ratio—64 dB,
- digital output type—PDM (Pulse-Density Modulation), and
- sensitivity—(–26 ± 3) dBFS.

In the sensor module, the microphone is connected directly to microcontroller via a digital interface SPI1. The whole of the measurement signal processing function is realized in the microcontroller by means of execution of dedicated software routines. The measuring module microcontroller is connected with the communication module via the use of SPI3 serial interface. Structure of the measuring module comprises type 24AA08 (Microchip) EEPROM memory used to store data on the measuring module. The data are read out by the communication module once the sensor module is connected. Electronic circuits of the measuring module were made in the form of two printed-circuit boards. One of the boards comprises a measuring microphone, whereas the remaining elements of the measuring module are mounted on the other. Such structure is a result of the necessity to situate the measuring microphone (which is mounted on the printed-circuit board with the use of the surface-mount technology)

FIGURE 4.27   The noise sensor module with its housing open and closed.

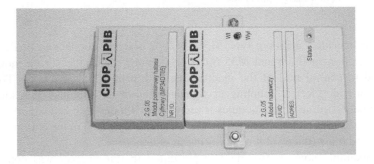

FIGURE 4.28   The noise measuring device formed as a result of connecting the noise sensor module (left) with a the communication module (right).

in a plane perpendicular to the main plane of the printed-circuit board. The housing for the measuring module was 3D-printed of a yellow-colored plastic, according to the color coding adopted throughout the network in order to facilitate differentiation between individual types of sensors (e.g., noise—yellow, mechanical vibrations—orange, etc). A view of the noise sensor module is presented in Figure 4.27.

Figure 4.28 shows a view of the noise measuring device which is created when the noise sensor module is being connected to the communication module.

The above-described modular structure can be applied to construct sensor modules for other harmful factors existing in the work environment and the developed sensor network can be used as an effective tool for reduction of workers' exposure to hazards from different measurable harmful factors.

4.5   SUMMARY

Examples of wireless sensor networks presented in this chapter included a system for remote supervision over correct usage of earmuffs and a sensor network for work environment monitoring and alerting workers to the hazard of excessive exposure noise. These examples show that solutions of these types can be deployed and used as effective tools supporting other measures of control of workers' exposure to noise and other hazards. The network components enable processing of data continuously and from many points at the same time which allows swift reaction to current hazards and offers a deeper insight into processes occurring in the environment. These merits of WSNs can be used for the purpose of carrying out more precisely oriented and more efficient preventive actions and projects aimed at improvement of safety and health conditions in the work environment. One may expect that further development of sensor networks, the Internet of Things, and artificial intelligence methods will result in creation of more human-friendly work environments actively protecting the worker against existing or newly emerging hazards. Specific challenges to achieving this goal include supplying power to sensor network elements guaranteeing sufficiently long periods of autonomous and maintenance-free operation. In this chapter, it has been shown that in any workplace there are at least several types of renewable energy sources which, even in the present state of the art, can be used to supply electric power to sensor networks monitoring the work environment.

REFERENCES

Ashton, K. 2009. That 'Internet of Things' thing. *RFID Journal*. http://www.rfidjournal.com/articles/view?4986. (accessed January 28, 2020).

Atzori, L., A. Iera, and G. Morabito. 2010. The Internet of Things: A survey. *Comput Netw* 54(18):2787–2805. DOI:10.1016/j.comnet.2010.05.010.

Berger, E. H. 1998. Can real-world hearing protector attenuation be estimated using laboratory data? *Sound Vib* 22(12):26–31.

Berger, E. H., L. H. Royster, J. D. Royster, D. P. Dricsoll, and M. Layne, eds. 2003. *The Noise Manual*. Revised 5th edition. Fairfax, VA: American Industrial Hygiene Association.

Brettel, M., N. Friederichsen, M. Keler, and M. Rosenberg. 2014. How virtualization, decentralization and network building change the manufacturing landscape: An industry 4.0 perspective. *Int J Inform Commun Eng* 8(1):37–44.

Burks, J. A., and R. R. Stein. 1998. Laboratory investigation of factors affecting the real world performance of earmuffs. American Industrial Hygiene Conference, San Francisco, CA, paper 42.

Canetto, P. 2009. Hearing protectors: Topicality and research needs. *Int J Occup Saf Ergon* 15(2):141–153. DOI:10.1080/10803548.2009.11076795.

Chabanne, H., P. Urien, and J.-F. Susini, eds. 2011. *RFID and the Internet of Things*. London, UK: ISTE; Hoboken, NJ: Wiley.

Dargie, W., and C. Poellabauer. 2010. *Fundamentals of Wireless Sensor Networks: Theory and Practice*. Chichester, UK; Hoboken, NJ: Wiley.

Delicato, F. C., P. F. Pires, and T. Batista. 2013. *Middleware Solutions for the Internet of Things*. London, UK: Springer.

Directive 2003/10/EC of the European Parliament and of the Council of 6 February 2003 on the minimum health and safety requirements regarding the exposure of workers to the risks arising from physical agents (noise) (Seventeenth individual Directive within the meaning of Article 16(1) of Directive 89/391/EEC). OJ 42/38, 15.2.2003.

Dziadak, B., Ł. Makowski, and A. Michalski. 2016. Survey of energy harvesting systems for wireless sensor networks in environmental monitoring. *Metrol Meas Syst* 23(4):495–512. DOI:10.1515/mms-2016-0053.

Galio, S., and G. Lo Re, eds. 2014. *Advances onto the Internet of Things. How Ontologies Make the Internet of Things*. Cham, Switzerland; New York: Springer-Verlag.

Gubbi, J., R. Buyya, S. Marusic, and M. Palaniswami. 2013. Internet of Things (IoT): A vision, architectural elements and future directions. *Future Gener Comp Syst* 29:1645–1660. DOI:10.1016/j.future.2013.01.010.

Kagermann, H., W. D. Lukas, and W. Wahlster. 2011. Industrie 4.0: Mit dem Internet der Dinge auf dem Weg zur 4. industriellen Revolution. http://www.vdi-nachrichten.com/Technik-Gesellschaft/Industrie-40-Mit-Internet-Dinge-Weg-4-industriellen-Revolution. (accessed January 28, 2020).

Karimi, K., and G. Atkinson. 2014. What the Internet of Things (IoT) needs to become reality. White Paper, freescale.com/arm.com, https://www.nxp.com/docs/en/white-paper/INTOTHNGSWP.pdf. (accessed January 28, 2020).

Koradecka, D., ed. 2010. *Handbook of Occupational Safety and Health*. Boca Raton, FL: CRC Press, Taylor & Francis Group.

Kotarbińska, E., E. Kozłowski, and W. Barwicz. 2007. Evaluation of individual exposure to noise when ear-muffs are worn. *Proceedings of First European Forum on Efficient Solutions for Managing Occupational Noise Risks*, Noise at work 2007, July 2–5, 2007, Lille, France.

Kotarbińska, E., and E. Kozłowski. 2009. Measurement of effective noise exposure of workers wearing ear-muffs. *Int J Occup Saf Ergon* 15(2):193–200.

Kritz, P., F. Maly, and T. Kozek. 2016. Improving indoor localization using bluetooth low energy beacons. *Mob Inf Syst* 2016:e2083094. DOI:10.1155/2016/2083094.

Ma, Z., S. Poslad, J. Bigham, X. Zhang, and L. Men. 2017. A BLE RSSI ranking based indoor positioning system for generic smartphones. *Proceedings of 2017 Wireless Telecommunications Symposium*, April 26–28, 2017, Chicago.

Miorandi, D., S. Sicari, F. De Pellegrini, and I. Chlamtac. 2012. Internet of things: Vision, applications and research challenges. *Ad Hoc Netw* 10(7):1497–1516.

Morzyński, L. 2013. Wireless monitoring system supporting correct use of earmuffs. *Proceedings of Inter-Noise 2013*, September 15–18, 2013, Innsbruck, Austria.

Morzyński, L. 2014. Zastosowanie modułów XBee do monitorowania narażenia na hałas pracowników użytkujących nauszniki przeciwhałasowe. *Przegląd Elektrotechniczny* 9:187–190. DOI:10.12915/pe.2014.09.47.

Morzyński, L. 2015. System zdalnego monitorowania parametrów wibroakustycznych z wykorzystaniem odnawialnych źródeł energii [Vibroacoustic parameters remote measurement system with the use of renewable energy sources]. *Bezpieczeństwo Pracy – Nauka i Praktyka* 11:13–17.

Morzyński, L. 2019a. Idea wykorzystania bezprzewodowej sieci sensorowej i Internetu rzeczy do monitorowania środowiska pracy i ostrzegania pracowników przed zagrożeniami [The idea of the use of wireless sensor networks and the internet of things for monitoring the working environment and warning workers about hazards]. *Bezpieczeństwo Pracy – Nauka i Praktyka* 1:23–26. DOI:10.5604/01.3001.0012.8511.

Morzynski, L. 2019b. IoT-based system for monitoring and limiting exposure to noise, vibration and other harmful factors in the working environment. *Proceedings of Inter-Noise 2019*, June 16–19, 2019, Madrid, Spain.

Morzyński, L., and G. Szczepański. 2017. Model systemu zdalnego monitoringu parametrów wibroakustycznych środowiska pracy [Model of the wireless monitoring of noise and vibration in the working environment's system]. *Bezpieczeństwo Pracy – Nauka i Praktyka* 4:6–20. DOI:10.5604/01.3001.0009.8780.

Murphy, W. J., and J. R. Franks. 2000. Franks evaluation of a real-world hearing protector fit-test system. *Spectrum Suppl* 1:17–18.

NIOSH [National Institute of Occupational Safety and Health]. 1998. Criteria for a Recommended Standard. Occupational Noise Exposure. Revised Criteria. Cincinnati, OH: NIOSH. https://www.cdc.gov/niosh/docs/98-126/pdfs/98-126.pdf. (accessed February 3, 2020).

Raghavendra, C. S., K. M. Sivalingam, and T. Znati, eds. 2006. *Wireless Sensor Networks.* New York: Springer.

Szczepański, G. 2017a. Wykorzystanie energii z ciepła odpadowego do zasilania sieci czujnikowych [The use of waste heat to power sensor networks]. *Rynek Energii* 5(123):60–65.

Szczepański, G., and L. Morzyński. 2017b. Zastosowanie techniki energy harvesting do zasilania sieci czujników bezprzewodowych na przykładzie systemu zdalnego monitoringu parametrów wibroakustycznych środowiska pracy [Use of energy harvesting technique to power wireless sensor network on example of system for wireless monitoring of vibroacoustic parameters of working environment]. *Wibroakustyka* 4:31–37. DOI:10.15199/148.2017.4.1.

Vergone, R., D. Dardari, G. Mazzini, and A. Conti. 2008. *Wireless Sensor and Actuator Network: Technologies, Analysis and Design.* London, UK: Academic Press.

Wang, C., Q. Zhang, and W. Wang. 2017. Low-frequency wideband vibration energy harvesting by using frequency up-conversion and quin-stable nonlinearity. *J Sound Vibr* 399:1691–1181. DOI:10.1016/j.jsv.2017.02.048.

Wortmann, F., and K. Fluchter. 2015. Internet of Things–A technology added. *Bus Inform Syst Eng* 57(3):221–224.

Yang, S.-H. 2014. *Wireless Sensor Networks: Principles, Design and Applications.* London, UK: Springer.

Yick, J., B. Mukherjee, and D. Ghosal. 2008. Wireless sensor network survey. *Comput Netw* 52(12):2292–2330. DOI:10.1016/j.comnet.2008.04.002.

5 Genetic Optimization Techniques in Reduction of Noise Hazards

Leszek Morzyński

CONTENTS

5.1 INTRODUCTION

5.1.1 THE GENETIC ALGORITHM

Effectiveness of technical measures taken to reduce noise depends on many factors—such as dimensions, mass, shape, or the sound absorption coefficient—which are usually strongly interrelated [Berger et al. 2003; Vér and Beranek 2006; Engel et al. 2010]. These parameters, together with specific constraints imposed on them, must be taken into account in the process of designing noise control measures. In many cases, it is possible to achieve the same minimization of noise-related hazards by employing different measures and methods. In such cases, the overall cost of deploying a specific solution in a given workplace becomes an important issue. It can therefore be assumed that the problem of reducing workers' exposure to noise is a multidimensional optimization problem consisting of a search for the best solution for minimizing the effect of noise on workers and depending on a specific set of parameters (variables) defining the problem. Among the most universally used classic optimization methods one can include analytical (gradient) techniques and enumeration methods (e.g., exhaustive search), each of which however have their specific disadvantages and limitations. Analytical methods require that the so-called objective function, or a function representing the quality of a solution dependant on the assumed set of variables, is continuous and differentiable. Analytical methods are also local by their nature (i.e., they consist in searching for the optimum solution in the neighborhood of a starting point defined in the space of variables). Therefore, analytical methods may lead to finding a local instead of the global solution to the problem. Effectiveness of the two above-described types of methods, especially enumeration ones, decreases as the number of variables defining the optimization problem increase. For that reason, another class of methods based on random search in space of variables (to which genetic algorithms belong) are enjoying increasing popularity among modern tools for solving optimization problems.

Genetic algorithms are optimization methods based on mechanisms governing biological evolution, learned from the observation of flora and fauna. The key mechanism of the theory of evolution formulated by Charles Darwin (1809–1882) is natural selection, according to which organism showing better fitness to a given natural environment increases their chances of survival and reproduction. The fact of "being fit" is determined by a set of favorable traits of given organisms. The traits are encoded in the organisms' genes and are subject to inheritance in the process of reproduction. The main source of variability is mutation, constituting a sudden step change in the genetic information record content. These mechanism are taken into consideration in genetic algorithms.

Genetic algorithms were invented by John Holland who published the theoretical foundations of the method in the 1960s [Holland 1962] and developed it further for some time afterward [Holland 1975]. Considerable progress in the area of genetic algorithms occurred at the turn of the 1980s and 1990s. Among the most important titles on genetic algorithms published during that period, one should mention classic books by Goldberg [1989] and Davis [1991]. The publishing market offers a large selection of more recent monographs devoted to theory and applications of genetic algorithms [Haupt and Haupt 1998; Bäck et al. 2000; Gwiazda 2006, 2007; Sivanandam and Deepa 2008].

The nomenclature used in the theory and practice of genetic algorithms is derived from biological sciences. The genetic algorithm operates on sets of coded variables of the optimization problem that needs to be solved. A coded optimization problem variable v is called a *gene*. A vector (set) of N genes makes up a *chromosome* **v**:

$$\mathbf{v} = \langle v_1, v_2, \ldots, v_N \rangle \tag{5.1}$$

A chromosome is therefore a coded representation of one possible solution to the problem which is called an *individual*. A set of chromosomes is called a *population*. For each chromosome, its fitness can be determined as the value of a specifically adopted fitness function (objective function) f the domain of which is the space of variables making up a chromosome:

$$f(\mathbf{v}) = f(v_1, v_2, \dots, v_N)$$ (5.2)

The fitness value allows one to evaluate whether a given solution to the problem (chromosome) is better or worse than other solutions. This information is necessary to orient the optimization process properly.

In the classical binary genetic algorithm, variables are coded in the form of binary sequences with length of L bits (Figure 5.1).

As a result of using binary representation and chromosomes with limited lengths, the binary genetic algorithm operates on a finite set of parameter values. In the case of dealing with parameters assuming continuous values, they must be subject to a quantization process before being coded. Other methods of coding can be also adopted in genetic algorithms, including coding with the use of real (floating-point) numbers which is used in the so-called continuous genetic algorithms [Haupt and Haupt 1998; Sivanandam and Deepa 2008].

An elementary genetic algorithm comprises three fundamental operations: selection, crossover, and mutation. A schematic diagram representing operation of a genetic algorithm is presented in Figure 5.2. The algorithm must include definitions of variables in terms of which the optimization problem is expressed, the chromosome structure, the fitness function, and any parameters of the algorithm used in individual stages of its operation (e.g., initial population size, selection method, crossover method, etc.). An algorithm starts from creating, randomly in most cases, an initial population of chromosomes, known also as the parent population. Next, the value of the fitness function is determined for each chromosome. In the selection process, certain chromosomes are selected to take part in the reproduction process. The higher the fitness of an individual characterized by given chromosome, the higher is the probability of that chromosome to be selected. The most commonly used selection methods are: the *proportionate selection* (or the roulette-wheel selection) in which the likelihood of selecting a chromosome for reproduction is proportional to its fitness, and the *tournament selection* in which couples (pairs) or larger subsets of chromosomes are randomly drawn from which the chromosome with the highest fitness is selected.

Pairs of individuals chosen in the course of selection are subject to the reproduction process which is effected by means of crossing over pairs of chromosomes (Figure 5.3). A crossover may be either a single-point or multi-point operation. Crossover points are selected randomly within a chromosome.

During crossover, new chromosomes (offspring) are created. Genes from new chromosomes are subjected to mutation which is an operation in which, with a low random probability, content of a selected gene is changed. As a result of crossover and mutation operations, a new generation of chromosomes, known also as the offspring or child population, comes into existence. In the next step of the algorithm,

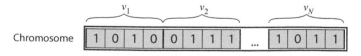

FIGURE 5.1 A chromosome with binary coding and code sequence length of $L = 4$ bits.

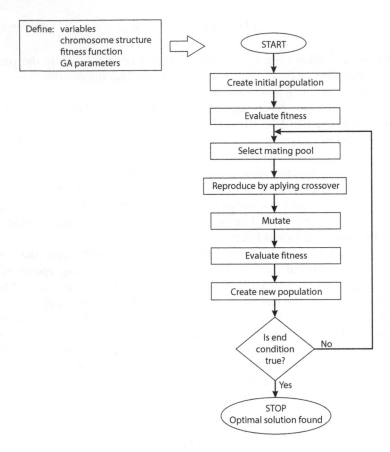

FIGURE 5.2 A genetic algorithm block diagram.

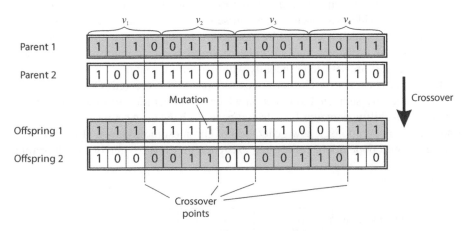

FIGURE 5.3 Multi-point crossover and mutation of binary chromosomes in case of chromosomes representing four variables.

the offspring population is used to create a new (parent) population. The simplest and most frequently used method consists in replacing the whole parent population with the offspring population. In more complex cases, individual chromosomes from the offspring population and the parent population are selected for the new population. Selection of best fit individuals from the parent population to enter the new population is called the elitist strategy or elitist selection. Operation of the genetic algorithm is continued until a termination condition is met. In the classical approach, such a termination condition consists in reaching a predefined number of generations (as counted from the algorithm operation start). Other conditions may provide for reaching a predefined fitness value or absence of changes in the average fitness value for a predefined number of successive generations.

The above review of the theory of genetic algorithms is only a brief introduction to the subject. Intensive research into the domain of genetic algorithms resulted in development of many different methods (and their variants) for creation of initial population, selection, crossover, mutation, and creation of new populations. Comprehensive discussion on the methods exceeds the scope of the present essay, but interested readers are hereby referred to a wide choice of literature on the subject [Goldberg 1989; Davis 1991; Haupt and Haupt 1998; Gwiazda 2006, 2007; Sivanandam and Deepa 2008].

Genetic algorithms, like other optimization algorithms, have their specific merits and limitations. Among the good points of genetic algorithms, one should include the versatility of the method which can be applied to many different types of optimization problems. The search for the optimum solution in a genetic algorithm is continued from many starting points simultaneously (initial population). Thanks to the parallel nature of the process of looking for the optimum solution and its orientation based solely on the fitness function, these methods are highly effective in finding global maxima even in the case of very complex optimization problems. On the other hand, one difficulty in application of genetic algorithms may consist in defining the fitness function and selection of individual parameters of the genetic algorithm deemed to have an essential effect on its operation. In some specific applications, the necessity to repeatedly evaluate the fitness function for chromosomes may become an issue. Moreover, genetic algorithms are metaheuristics which means that within a reasonable time frame, a quality solution can be usually found only to some accuracy, but finding the accurate global optimum may require additional application of other methods such as local search or gradient techniques.

5.1.2 APPLICATIONS OF GENETIC ALGORITHMS IN NOISE HAZARD CONTROL

Genetic algorithms are widely used in many disciplines of science and technology where multi-parameter optimization problems need to be resolved. In the following paragraphs, based on literature of the subject, a short review of application of genetic algorithms in acoustics is presented, with emphasis put on noise control.

Noise barriers are counted among most basic technical means of noise reduction. In Baulac et al. [2008], genetic algorithms were used to optimize the form of the upper surface of a T-shaped noise barrier with the objective of achieving the highest possible effectiveness of the barrier. Similarly, in Baulac et al. [2007], the shape of a sound screen with multiple upper edges was optimized; whereas in

Grubeša et al. [2012], the sound barrier structure optimization process accounted for not only the effectiveness of the device but also the material and form of barrier components as well as the resulting cost of realization of the whole project.

Other well-known technical measures of noise reduction are noise suppressors such as mufflers in combustion engine fume exhaust system outlets or heating, ventilation, and air conditioning (HVAC) systems. In Chang et al. [2005], the genetic algorithm was used to optimize parameters of a single-chamber noise muffler with constraints connected with its overall dimensions taken into account. Similar application of genetic algorithms for a multi-chamber muffler problem was presented in Chiu [2010]. An example of application of the genetic algorithm to optimization of parameters of an air compressor outlet muffler was given in Chang et al. [2017].

An example of new trends in noise reduction which enjoy increasing popularity in recent years are acoustic metamaterials [Zangeneh-Nejad and Fleury 2019]. Due to their complex internal structure, these materials are characterized by sound-insulating properties better than those predicted based on the grounds of the mass law or the individual properties of component materials used to construct them. The process of developing such materials is typically a process of multi-parameter optimization in which genetic algorithms may be successfully employed. Examples of application of genetic algorithms in development of new acoustic metamaterials are presented in Blevins [2016] and Lagarrigue et al. [2016].

In active noise control systems, genetic algorithms are used to optimize parameters of the system controller [Makarewicz 2007] and arrangement of measuring transducers and actuators [Aslan and Paurobally 2016].

Genetic algorithms can also be used to solve problems concerning noise level prediction. An example of such application is presented in Gundogu et al. [2005]. In the paper, a genetic algorithm is used in a road traffic noise prediction study. Some reduction of road traffic noise can be achieved by, for instance, optimization of geometrical parameters of slowing bumps determined with the use of the genetic algorithm [Sarafraz et al. 2016].

An important area of application of genetic algorithms in acoustics pertains to the problem of identification of sound sources. In Lan and Chiu [2008], the genetic algorithm was used for identification and determination of location of main noise sources on a factory site. Liu [2016] presents an application of genetic algorithms to localization of partial discharges in power transformers evidencing deterioration of parameters of electric insulation in the devices. Le Courtois et al. [2016] reports the possibility of applying the genetic algorithm method to optimization of a microphone array aimed at improvement of determining the location of moving sound sources with the use of the beamforming method. Performance of the optimized matrix was then validated in tests concerning noise sources generated by high-speed trains (TGV).

Genetic algorithms can also be used to optimize strategies and measures of noise control and reduction of exposure to noise in enterprises. Waly and Sarker [1998] present application of the genetic algorithm method to optimization of arrangement of noise sources (machines) in an industrial plant where the goal was to obtain the maximum permissible exposure time at a given point in space (the time interval for which the permissible value of exposure to noise will not be exceeded). In the optimization task, the issues taken into account included properties of the room in which noise sources were situated, directivity of the sources, and their dimensions together with the resulting constraints concerning their deployment. In Asawarungsaengkul and Nanthavanij [2006], application of genetic algorithms is presented to solve the problem of finding an optimal noise-related hazard reduction strategy in a work environment with a limited budget available for realization of necessary measures. The proposed procedure of selecting an optimum strategy for noise hazard control includes thirteen steps in the course of which, six models are used concerning optimization of the use of technical noise control measures, job rotation, and the use of hearing protectors. The problem of optimization of usage of resources available for noise reduction under limited budget conditions is also raised in Davis [2009]. An example reported in the paper concerned minimization of the cost of a project aimed at reduction of combined sound power level of all of 100 noise sources in an industrial plant with specifically determined parameters by 6.04 dB. The noise reduction occurred as a result of application of technical measures the price of which depend on the insertion loss and noise source sound power. Similar application of the genetic algorithm technique is also described in Razavi et al. [2014]. In the process of optimizing the strategy used in the scope of reduction of exposure to noise, the parameters taken into account included constraints concerning the total cost of the undertaking and the exposure period at which the exposure to noise for each worker did not exceed an permissible value.

In the next section, the genetic algorithm implementation process is described in more detail using the example of a computer program (application) for optimization of the arrangement of noise sources and workstations oriented at minimization of exposure to noise, including discussion of performance of the algorithm depending on parameters adopted for the optimization problem and the corresponding algorithm.

5.2 OPTIMIZATION OF SOUND SOURCE AND WORKSTATION ARRANGEMENTS ORIENTED TOWARD THE MINIMIZATION OF WORKERS' EXPOSURE TO NOISE

5.2.1 THE OPTIMIZATION PROBLEM

Proper arrangement of machines and equipment (noise sources) and workstations in a workplace can be counted among most basic methods of reduction of occupational exposure to noise. At a workstation, the A-weighted sound pressure level $L_{p,A}$ of noise generated by a given noise source (a noisy machine) depends on the A-weighted sound power level $L_{W,A}$ of the source and on the distance r between the workstation and the source. In cases when there are more than one noise source in the workstation's surrounding, each of the sources contributes to the resultant A-weighted sound pressure level observed at the workstation. By changing the positions of workstations relative to noise sources it is possible to affect the A-weighted sound pressure level values at workstations and thus also the workers' exposure to noise at their workstations. Therefore, the optimization problem consists in general in finding an arrangement of workstations and noise sources, with constraints concerning their relative positions taken into account, for which exposure to noise in workers present in the room will be as low as possible. The problem may be narrowed to searching for locations of machines only (if, for instance, arrangement of workstations is fixed or strictly attached to location of machines) or finding optimum arrangement of workstations only (where, e.g., locations of machines is fixed in view of a specific production line layout). The constraints concerning location of machines and workstations can be of different complexities, from the simplest ones (connected with overall dimensions of machines and workstations) to more complex restrictions that might result from requirements concerning relative positions of machines making up a specific production line. In the algorithm presented later in this chapter, the computer program is an embodiment of the idea of the algorithm and acts as a demonstration of its potential and properties. The dimension constraints mentioned above are used in that example. Constraints of that type were defined in the already quoted paper by Waly and Sarker [1998] concerning optimization of location of noise sources aimed at reduction of workers' exposure to noise.

5.2.2 Chromosome Structure, Fitness Function, and Genetic Algorithm Parameters

In the considered problem, optimal location of noise sources and workstations which results in minimal workers' exposure to noise should be found [Morzyński 2010a, 2010b, 2011]. Therefore, coordinates of the noise sources and workstations in three-dimensional space are the variables of the optimization problem. The structure of the proposed chromosome is shown in Figure 5.4. The first section of the chromosome consists of coordinates of sound sources and the second represents coordinates of workstations. If the optimization problem concerns M noise sources and N workstations, a chromosome consist of $3 \times (N + M)$ variables representing spatial coordinates.

It was assumed that the genetic algorithm used to optimize arrangement of noise sources and workstations would be a binary algorithm. In view of the above, the value of each of the position coordinates x, y, and z represented by real numbers needs to be coded as a binary sequence with definite length of L_{bit} bits. For obvious reasons, conversion of real numbers in floating point representation to a L_{bit}-long binary number results in quantization of the value of the quantity. The value of the coded variable v_{bin} can be obtained by determining the integer part of the product of the unified variable (normalized with respect to the range of the variable) and the maximum number which can be converted into L_{bit}-long binary number, according to the formula [Lan and Chiu 2008]

$$v_{bin} = \text{int}\left\{ \frac{v_{real} - v_{min}}{v_{max} - v_{min}} \left(2^{L_{bit}} - 1\right) \right\} \tag{5.3}$$

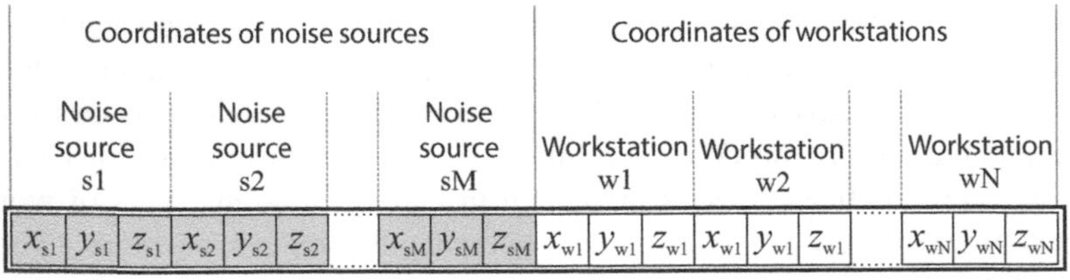

FIGURE 5.4 Chromosome structure in the considered optimization problem concerning arrangement of noise sources and workstations.

where:

v_{bin}: variable value (coordinate x, y, or z) after coding
v_{real}: real variable value
v_{min}: minimum variable value
v_{max}: maximum variable value
L_{bit}: encoding string length

The variable is decoded according to the formula

$$v_{real} = v_{bin} \frac{v_{max} - v_{min}}{2^{L_{bit}} - 1} + v_{min} \tag{5.4}$$

The above described method of coding parameters allows setting some constraints for both the genetic algorithm and the achieved results. By setting maximal and minimal values of coordinate, it can be assured that achieved results are always in the given range. Setting the length of the binary string allows establishing the accuracy of the optimization process and the grid in which noise sources and workstations can be placed. The total length of the vector (chromosome) with a binary coded set of parameters defining positions of sound sources is $3 \cdot L_{bit} \cdot (N + M)$ bits.

An important step in the genetic algorithm development process is the choice of the fitness function. In the analyzed example, the function should be selected in a way enabling assessment of the results of operation of the algorithm in the scope of reduction of worker's exposure to noise. The simplest way of doing that consists in comparing values of the noise exposure levels ($L_{EX,8\,h}$) at individual workstations with the maximum permissible value (the value adopted in many countries and recommended by, among others, by the US National Institute for Occupational Safety and Health (NIOSH) [NIOSH 1998] is 85 dB). Under the assumption that the exposure time is eight hours, it is possible to compare the A-weighted sound pressure at workstations with the value of A-weighted pressure corresponding to the A-weighted sound pressure level of 85 dB. The formula defining the fitness function can be therefore given the following form:

$$f(\mathbf{v}) = \sum_{i=1}^{N} \left[\frac{10^{0.1 \cdot 85}}{10^{0.1 \cdot L_{p,A,i}^{(\mathbf{v})}}} \right] \tag{5.5}$$

where $\mathbf{v}$ is a chromosome, N is the number of workstations, and $L_{p,A,i}^{(\mathbf{v})}$ is the resultant A-weighted sound pressure level at i-th workstation for locations of the workstation and noise sources as coded in chromosome $\mathbf{v}$. The fitness value calculated in accordance with Equation 5.5 is the higher value, the lower value is exposure to noise in the whole of working staff. However, the fitness function defined that way is not free from some inconveniences. It can lead to solutions in which some workstations are "sacrificed"; in other words, they are exposed to high-level noise to minimize exposure to noise at other workstations and maintain chromosome fitness. To avoid situations of that kind, the fitness function should be defined in a way promoting distinct solutions in which exposure to noise at each workstation is kept below the maximum acceptable value. Such improved fitness functions may be defined, for instance, as follows:

$$f_e(\mathbf{v}) = \sum_{i=1}^{N} \left[\frac{10^{0.1 \cdot 85}}{\mu \cdot 10^{0.1 \cdot L_{p,A,i}^{(v)}}} \right], \quad \mu = \begin{cases} 1, & L_{p,A,i}^{(v)} \leq 80 \text{ dB} \\ 2, & 80 \text{ dB} < L_{p,A,i}^{(v)} \leq 85 \text{ dB} \\ 5, & L_{p,A,i}^{(v)} > 85 \text{ dB} \end{cases} \quad (5.6)$$

where μ is a weighting coefficient.

In this formula, a weighting coefficient was introduced which influences fitness value according to upper and lower exposure action values (as defined in Directive 2003/10/EC). A genetic algorithm using a fitness function defined by Equation 5.6 is expected to find the location of workplaces and noise sources, in which none of the workers is exposed to noise exceeding the exposure limit value and if possible, all of them are exposed to noise below the exposure action value. To calculate fitness of chromosome $\mathbf{v}$, it should be decoded and calculation of A-weighted sound pressure levels for each workstation should be performed using one of the well-known acoustics methods.

Optimization carried out with the usage of a genetic algorithm based on fitness functions as defined by Equation 5.6 or Equation 5.7 will not take into account the constraints of a real work environment connected, for example, with minimal distances between components of work environment (workstations, noise sources, work room walls). It is not possible for a workstation and a machine to be situated at the same point in space, although such solution will be eliminated from the population by the genetic algorithm itself due to low fitness value. However, a work environment can impose other constraints connected with positions of noise sources and workstations, not connected directly with the noise exposure value. Examples of such constraints include relative positions of workstations which, when situated very close to each other, admittedly contribute to reduction of exposure to noise but prevent the operators from carrying out effective work at their workstations. The problem of optimization of positions occupied by individual noise sources and workstations aimed at minimization of exposure to noise is therefore the optimization problem with constraints. One method applied commonly in genetic algorithms developed to solve optimization problems with constraints consists in using a so-called penalty function $p(\mathbf{v})$ [Yeniay 2005]. Penalty functions can modify the fitness function in either an additive or multiplicative manner. In the example discussed here, the multiplicative approach was adopted, in which [Morzyński 2010a, 2011]:

$$f_e(\mathbf{v}) = f_c(\mathbf{v}) p(\mathbf{v}) \quad (5.7)$$

where $f_c(\mathbf{v})$ is the expanded fitness function in which constraints specific for the optimization problem are represented by the function $p(\mathbf{v})$. In the following, examples are given of penalty functions applied in the algorithm enabling inclusion of specific constraints concerning minimum allowable distances between the noise sources and workstations in the process of optimization of the location of these elements.

To take into account the constraints concerning the minimum allowable distance between noise sources in the optimization process, the following penalty function was used:

$$p^{ss}(\mathbf{v}) = \prod_{j=1}^{M-1} \prod_{n=j+1}^{M} p_{j,n}^{ss}(\mathbf{v}) \tag{5.8}$$

where:

$$p_{j,n}^{ss}(\mathbf{v}) = \begin{cases} 1, & r_{j,n} \geq r_{\min,ss} \\ \dfrac{r_{j,n}^2}{r_{\min,ss}^2}, & r_{j,n} < r_{\min,ss} \end{cases} \tag{5.9}$$

is the penalty function for noise sources j and n, $r_{j,n}$ is the distance between noise sources j and n, and $r_{\min,ss}$ is the minimum distance between any two sources.

Constraints connected with the minimum allowable distance between workstations were taken into account by using the penalty function $p^{ww}(\mathbf{v})$ expressed by the formula

$$p^{ww}(\mathbf{v}) = \prod_{i=1}^{N-1} \prod_{m=i+1}^{N} p_{i,m}^{ww}(\mathbf{v}) \tag{5.10}$$

where:

$$p_{i,m}^{ww}(\mathbf{v}) = \begin{cases} 1, & r_{i,m} \geq r_{\min,ww} \\ \dfrac{r_{i,m}^2}{r_{\min,ww}^2}, & r_{i,m} < r_{\min,ww} \end{cases} \tag{5.11}$$

where $p_{i,m}^{ww}(\mathbf{v})$ is the penalty function for workstations i and m, $r_{i,m}$ is the distance between workstations i and m, and $r_{\min,ww}$ is the minimum allowable distance between any two workstations.

The algorithm may also provide for a constraint concerning the minimum distance between a noise source and a workstation by using a penalty function $p^{sw}(\mathbf{v})$ given by the formula

$$p^{sw}(\mathbf{v}) = \prod_{j=1}^{M} \prod_{i=1}^{N} p_{j,i}^{sw}(\mathbf{v}) \tag{5.12}$$

where:

$$p_{j,i}^{sw}(\mathbf{v}) = \begin{cases} 1, & r_{j,i} \geq r_{\min,sw} \\ \dfrac{r_{j,i}^2}{r_{\min,sw}^2}, & r_{j,i} < r_{\min,sw} \end{cases} \tag{5.13}$$

is the penalty function for noise source j and workstation i, $r_{j,i}$ is the distance between noise source j and workstation i, and $r_{\text{min,sw}}$ is the minimum allowable distance between any noise source and any workstation.

The expanded fitness function of a genetic algorithm (taking into account the above-described constraints of the optimization problem) can be finally written down as follows:

$$f_{\text{e}}\left(\mathbf{v}\right) = f_{\text{c}}\left(\mathbf{v}\right) \cdot p^{\text{ss}}\left(\mathbf{v}\right) \cdot p^{\text{ww}}\left(\mathbf{v}\right) \cdot p^{\text{sw}}\left(\mathbf{v}\right) \qquad (5.14)$$

Constraints resulting from minimum allowable distances between noise sources and/or workstations and individual surfaces defining the working space (i.e., walls, floor, and ceiling) can also be taken into account in the algorithm by properly defined penalty function. However, in the case discussed here, the constraints were taken into account directly in the process of coding the optimization problem variables. If the minimum allowable distance of a noise source or a workstation from the workroom surface is $r_{\text{min,room}}$, the constraint can be taken into account in the coding process by modifying the formula of Equation 5.3 and giving it the form

$$v_{\text{bin}} = \text{int}\left\{\frac{v_{\text{real}} - \left(v_{\text{min}} - r_{\text{min,room}}\right)}{\left(v_{\text{max}} - r_{\text{min,room}}\right) - \left(v_{\text{min}} - r_{\text{min,room}}\right)}\left(2^{L_{\text{bit}}} - 1\right)\right\} \qquad (5.15)$$

It should be noted that the discussed method of taking into account possible constraints of the optimization process in the genetic algorithm differs fundamentally from methods in which penalty functions are used, as it does not allow the algorithm to search for solutions in areas where the minimum distance condition is not met (corresponding values will not be coded in chromosome). Quite the opposite, when the penalty function is used, the algorithm allows for existence of solutions in areas where the distance criterion is not met but such solutions are eliminated in the course of algorithm operation in view of low fitness value they represent.

The constraints described above define a simplified work environment model. In actual work environments, constraints concerning possible locations of noise-emitting machines and workstations are typically much more complex than those resulting from minimum allowable distances between them. Workstations are usually ascribed to specific machines, locations of machines relative to each other follow from their role in a specific production cycle, etc. However, even a work environment model as simple as this one is sufficient to demonstrate the potential of the genetic algorithm in the scope of environment optimization relative to minimization of exposure to noise and to draw general conclusions concerning optimal arrangement of machines within a workroom area.

The following assumptions were further adopted for the developed genetic algorithm:

1. The initial population is generated based on the value of random variables with uniform distribution (RND). The initial population size L_{pop} is determined arbitrarily. The algorithm of creating the initial population of solution vectors (chromosomes) $\mathbf{v}$ can be presented symbolically in the form of the following pseudocode:

```
constant Lpop                        // population size
constant Lbit                        // code string length
constant N                           // number of workstations
constant M                           // number of noise sources
variable k                           // chromosome index
variable l                           // variable index
variable v_bin Lpop,3*(N+M)]         // array of variables in
chromosomes

for k = 1 to k = Lpop do {           // for each chromosome
    for l = 1 to l = 3*(N+M) do {    // for each variable
        v_bin[k,l] = int(RND*(2^Lbit-1))    // generate a random value
    }
}
```

2. Selection of individuals allowed to take part in crossover is carried out according to the proportional selection principle (roulette). The probability of an individual (chromosome) $\mathbf{v}_k$ to be selected for the parent population $\mathbf{v}_{k,\text{parent}}$ is proportional to its fitness; in other words, the probability is proportional to the value of the fitness function $f_e(\mathbf{v})$. In the form of a pseudo-code, the selection procedure can be outlined as follows:

```
constant Lpop                        // population size
variable k, l                        // chromosome indexes
variable v[Lpop]                     // array of chromosomes
variable pop_fitness = 0             // overall population fitness
variable sum                         // fitness partial sum
variable beta                        // random variable

for k = 1 to k = Lpop do {           // calculate overall population
                                        fitness
    pop_fitness = pop_fitness + fitness(v[k])
}
for k = 1 to k = Lpop do {           // continue for the whole
                                        population
    l = 1
    sum = fitness(v[l])/pop_fitness  // determine partial sum
                                     // for the first chromosome
    beta = RND                       // generate a random number
    while sum < beta do {            // add fitness partial sums
        l = l + 1                    // until the sum reaches beta
        sum = sum + fitness(v[l]/pop_fitness)
    }
    v[k] = v[l]                      // select k-th parent
}
```

3. Multi-point crossover—the number of crossover points equals the number of variables (spatial coordinates) codes in a chromosome and the crossover procedure assures that within the scope of a single variable, there is zero or one crossover point. The crossover point (cross_point) within a gene is selected randomly (with the same probability). The crossover occurs with a certain probability P_{crs}. Symbolically, the crossover procedure can be written down as follows:

```
constant Lpop                      // population size
constant Lbit                      // code string length
constant N                         // number of workstations
constant M                         // number of noise sources
constant Pcrs                      // crossover probability
variable k                         // chromosome index
variable l                         // variable index
variable beta, alpha               // random numbers
variable cross_point               // crossover point (bit number)
                                   // within the variable (gene)
variable v_parent[Lpop,3*(N+M)]    // array of variables
                                       // in parent chromosomes
variable v_offspring[Lpop,3*(N+M)] // array of variables in offspring
                                   //chromosomes
function bits(v_bin, b, a)         // a function selecting a sequence
                                   // of bits numbered from a (LSB)
                                   //to b (MSB) from variable v_bin
for k = 1 to k = Lpop {            // continue for the whole population
    for l = 1 to l = 3*(N + M) {   // for each variable
    beta = RND                     // generate random number beta
    if Pcrs > beta then {          //if crossover probability > beta then
                                   //cross over within variable l
        alpha = RND                // generate random number alpha
        cross_point = int[alpha*(lbit-1)] // determine crossover point
                                   // cross over within variable l
        v_offspring[k,l] = (bits(v_parent[k,l], Lbit ,cross_point),
                        bits(v_parent[k+1,l], cross_point-1, 1))
        v_offspring[k+1,l] = (bits(v_parent[k+1,l], Lbit ,cross_point),
                        bits(v_parent[k,l], cross_point-1, 1))
    }
    else {                         // if crossover probability =< beta
                                   //then copy variables unchanged
        v_offspring[k,l] = v_parent[k,l]
        v_offspring[k+1,l] = v_parent[k+1,l]
    }
    }
k = k + 2
}
```

4. Uniformity of the probability P_{mut} of mutation within a chromosome. Symbolically,

```
constant Npop                        // population size
constant Lbit                        // code string length
constant N                           // number of workstations
constant M                           // number of noise sources
constant Pmut                        // mutation probability
variable beta                        // random number
variable k                           // chromosome index
variable n_bit                       // bit index in chromosome
variable v[Npop]                     // array of chromosomes


for k = 1 to k = Npop {              // for each chromosome and each bit
    for n_bit = 1 to n_bit = 3*(N + M)*Lbit {
    beta = RND                       // generate a random number beta
                                     // if mutation probability > beta
    if Pmut > beta then v[k] (n_bit)= ~v[k](n_bit)  // flip bit no. n_bit
    }
}
```

5. Application of elitist strategy in the process of selecting and creating a new population by choosing *E* best chromosomes from among the parent generation to a new generation. Symbolically,

```
constant Npop                        // population size
constant E                           // elitist population size
variable v_new_pop[Npop]             // array of new population
                                        chromosomes
variable v_parents[Npop]             // array of parent chromosomes
variable v_offsprings[Npop]          // array offspring chromosomes
variable k                           // chromosome index

for k = 1 to k = E {                 // copy chromosomes
    v_new_pop[k] = v_ parents[k]        // from elite parent population
}
for k = E+1 to k = Npop {            // continue copying chromosomes
    v_new_pop[k] = v_     offsprings[k-M]   // from offspring population
}
```

6. Termination of the optimization process when a predefined arbitrarily chosen number of generations is reached (i.e., after a determined number of algorithm iterations) L_{gnr}. In the form of a pseudocode,

```
constant Lgnr                        // number of generations
variable g = 1                       // generation index
variable v                           // population of chromosomes


while k < Lgnr do {
    select_mating_pool(v)
    reproduce_by_crossover(v)
    mutate(v)
    evaluate_fitness(v)
    create_new_population(v)
}
```

5.2.3   A Genetic Algorithm-Based Software Application
for Noise Sources and Workstations Arrangement

Based on the described genetic algorithm, a computer application for research purposes was developed (described earlier in [Morzyński 2010a, 2011]). This application has two modes of operation: design and calculation. In the design mode (Figure 5.5), a model of an acoustic work environment can be created. This model consists of a work room, workstations, and noise sources in an initial configuration (location). The work room dimensions and the number as well as initial positions of noise sources (marked as circles with letter "s" and a number in Figure 5.5) and workstations (marked as squared with letter "w" and a number) can be entered by the user via a graphical interface available in the application. All of these work environment elements can be changed and edited.

On the right side of the application window there is the variable and parameter list, in which work environment as well as genetic algorithm parameters can be

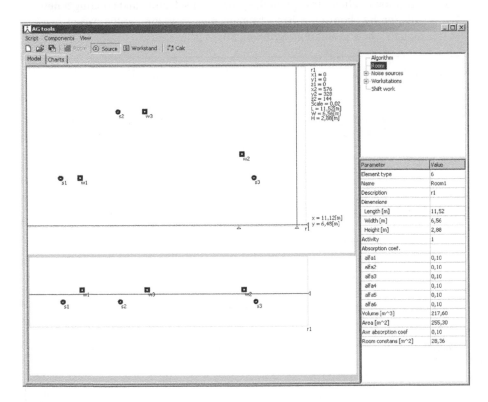

FIGURE 5.5   Computer application—the work room design mode.

set or changed. The work room is characterized by dimensions and the absorption coefficient of each wall. Noise sources are characterized by location coordinates, A-weighted sound power level, and status. Workplaces are characterized by location coordinates and status. Status is represented by a variable which affects calculation and optimization procedure. It allows noise sources and workplaces not to be considered in calculations and/or optimization simulations (e.g., depending on status noise source will not be considered in calculation of A-weighted sound pressure level or its location will stay unchanged during the optimization process). The genetic algorithm parameters which can be user-defined in the application are: population size, code string length, crossover and mutation probability, number of algorithm generations, and minimum distances for the penalty function. Algorithm options also provide for the choice of which of the position-defining variables (z, y, or z) shall be coded in the genetic algorithm chromosome. If, for instance, variable z is excluded from coding, then the genetic algorithm will have no effect on the position of any noise source or workstation along z-axis; in other words, the height at which the object is situated will not change. The selection method used in the application was the elitist strategy in which one chromosome from the parent population is advanced to the new generation ($E = 1$). Another parameter of the application denoted *algorithm_type* offers the choice whether the optimization applies to location of workstations only, noise sources only, or sound sources and workstations at the same time.

In calculation mode, the distribution of A-weighted sound pressure levels in a workroom (Figure 5.6) as well as occupational risk areas (Figure 5.7) can be determined and presented. Next, optimization can be performed with the use of the genetic algorithm. Calculations of the A-weighted sound pressure level at individual points of the work environment (and in particular at workstations) are necessary for determining the fitness function values according to Equation 5.6 and to enable the application to visualize results of operation of the algorithm. The application displays the results in the form of two-dimensional maps of the A-weighted sound pressure level distribution pattern on a work room cross section at a selected height (Figure 5.6). Additionally, in a window below the map, a plot of A-weighted sound pressure level variation along a selected straight line of the plane is presented.

The A-weighted sound pressure level at a workstation (or at any other point of the workroom) is a result of the combined effect of individual noise sources. Its value is determined according to the formula

$$L_{p,A,i} = 10 \log_{10} \left(\sum_{j=1}^{M} 10^{0.1 \cdot L_{p,A,i,j}} \right) \text{dB} \qquad (5.16)$$

in which i denotes workstation number, j is the noise source number, and $L_{p,A,i,j}$ is the A-weighted sound pressure level at workstation i resulting from emission of source j.

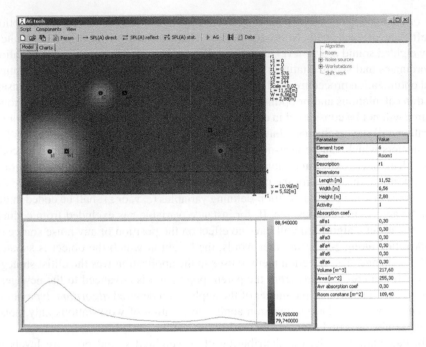

FIGURE 5.6 Computer application in calculation mode—distribution of A-weighted sound pressure level.

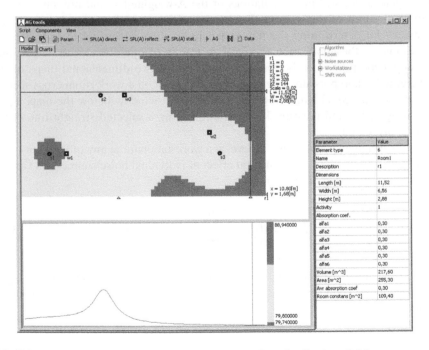

FIGURE 5.7 Computer application in calculation mode—distribution of risk areas.

Taking into account only the direct sound wave (with the room acoustics neglected) and assuming that all the sources are characterized by the omnidirectional radiation directivity pattern, the A-weighted sound pressure level $L_{p,A,i,j}$ representing the effect of j-th sound source as i-th workstation can be determined from the formula [Berger et al. 2003; Vér and Beranek 2006; Peters et al. 2011]

$$L_{p,A,i,j} = L_{W,A,j} - 20\log_{10} r_{i,j} - 10\log_{10}\left(4\pi\right) \text{ dB} \qquad (5.17)$$

where $L_{W,A,j}$ is the A-weighted sound power level of the j-th source and $r_{i,j}$ is the distance between i-th workstation and j-th source.

With the directivity of sound source radiation and the effect of room boundaries (reverberant sound) taken into account, the A-weighted sound pressure level can be determined from the formula derived on the grounds of Sabine's statistical theory of room acoustics [Berger et al. 2003; Vér and Beranek 2006; Peters et al. 2011]:

$$L_{p,A,i,j} = L_{W,A,j} + 10\log_{10}\left[\frac{Q_j}{4\,r_{i,j}^2} + \frac{4}{R}\right] \text{ dB} \qquad (5.18)$$

where $L_{W,A,j}$ is the A-weighted sound power level of the j-th source, Q_j is the directivity factor of the j-th source, $r_{i,j}$ is the distance between i-th workstation and j-th source, and R is the room constant calculated from the formula

$$R = \frac{S \cdot \bar{\alpha}}{1 - \bar{\alpha}} \qquad (5.19)$$

where S is the total surface area of the room and α is the average sound absorption coefficient of the work room surfaces.

The distance $r_{i,j}$ is given by

$$r_{i,j} = \sqrt{\left(x_i - x_j\right)^2 + \left(y_i - y_j\right)^2 + \left(z_i - z_j\right)^2} \qquad (5.20)$$

where x_i, y_i, z_i are coordinates of the i-th workstation, and x_j, y_j, z_j are coordinates of the j-th source.

In the described application, an assumption was adopted that all noise sources are omnidirectional point sources ($Q_j = 1$).

In the calculation mode it is possible, at any time, to switch from a view presenting A-weighted sound pressure levels to a view showing spatial distribution of occupational risk areas (Figure 5.7). The occupational risk areas depict the hazard concerning the level of exposure to noise ($L_{EX,8h}$) to which a worker staying in given area is subject. The risk can be low ($L_{EX,8h} \leq 80$ dB, dark gray shading), medium (80 dB $< L_{EX,8h} \leq 85$ dB, gray shading), or high ($L_{EX,8h} > 85$ dB, light gray shading).

When launched in the calculation mode aimed at optimization of the arrangement of noise sources and workstations, the application displays evolution of positions of the elements in consecutive generations of chromosomes (for the best fit) and the resulting changes in the A-weighted sound pressure level values or occupational risk (exposure) values. After completion of the optimization process, the arrangement of noise sources and workstations represented by the best fit chromosome in the last generation is displayed.

Based on the presented application, the effect of some genetic algorithm parameters and the penalty function on the algorithm operation is discussed. An example is also given using the application to optimization of noise source locations with an analysis of the effect of algorithm parameters on results of the optimization process.

5.2.4 THE EFFECT OF SELECTED GENETIC ALGORITHM PARAMETERS AND THE PENALTY FUNCTION ON GENETIC ALGORITHM OPERATION AND OPTIMIZATION RESULTS

In this subsection, the effect of selected parameters of the genetic algorithm and the penalty function on operation of the algorithm and the obtained results of the optimization process is discussed based on the example of simulations realized with the use of the computer application described in Section 5.2.3. The adopted model of an acoustic work environment was a workroom with the length of 52 m, the width of 31.2 m, the height of 6.4 m, and the average sound absorption coefficient of 0.55. In the workroom, eight noise sources (denoted s1–s8) and eight workstations (denoted w1–w8) were situated. The initial arrangement of noise sources and workstations and the resulting A-weighted sound pressure level distribution patterns and occupational risk areas are presented in Figures 5.8 and 5.9. The plots may be used as reference images to which results of the optimization process will be compared. The basic parameters of the genetic algorithm were as follows:

- generation size $L_{gnr} = 100$,
- population size $L_{pop} = 20$,
- crossover probability $P_{crs} = 0.6$, and
- mutation probability $P_{mut} = 0.001$.

It was assumed that workstations were located at the height of 1 m (the term "workstation height" should be understood as the position of a point at which the A-weighted sound pressure level is determined for the purpose of the noise exposure assessment and chromosome fitness evaluation). The only initial constraint to the optimization problem was the minimum distance between a noise source or a workstation and any of the surfaces defining the room volume, $r_{min,room} = 0.5$ m.

Example 1—The effect of distance between machines on algorithm operation and optimization result on example of noise sources arrangement optimization

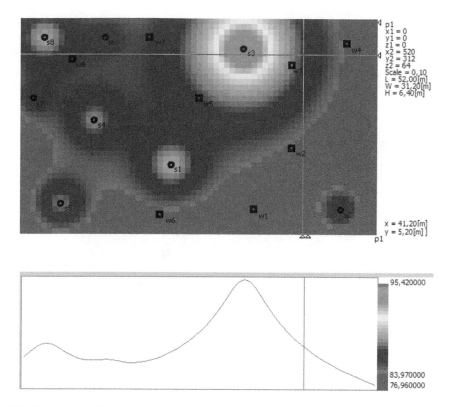

FIGURE 5.8   The initial arrangement of work environment elements and A-weighted sound pressure level distribution pattern on a room cross section at the assumed workstation height (a reference plot).

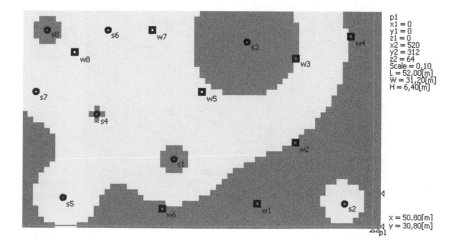

FIGURE 5.9   Initial distribution of the occupational risk area on a room cross section at the assumed workstation height (a reference plot).

Figures 5.10 through 5.12 illustrate the effect of the minimum allowable distance
between machines on the genetic algorithm operation results. For small minimum
allowable distances between machines (0.5 m—Figure 5.10), the effect of the penalty function is insignificant which results from the fact that with the condition
concerning the minimum allowable distance between machines being met, it is still

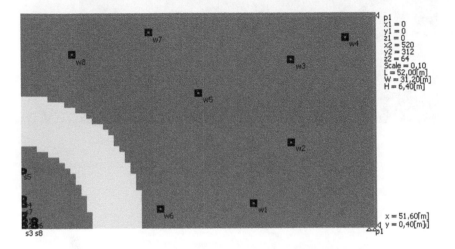

FIGURE 5.10 Arrangement of work environment elements and the occupational risk distribution pattern obtained as a result of operation of the genetic algorithm for the minimum
allowable machine–machine distance of 0.5 m.

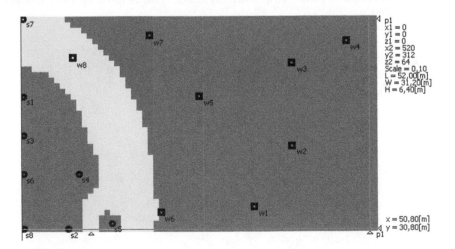

FIGURE 5.11 Arrangement of work environment elements and the occupational risk distribution pattern obtained as a result of operation of the genetic algorithm for the minimum
allowable machine–machine distance of 5 m.

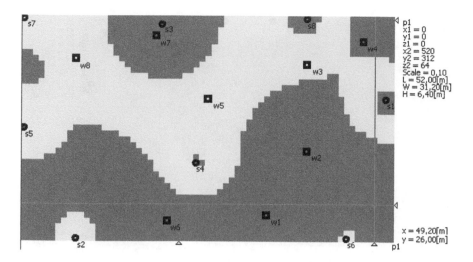

FIGURE 5.12 Arrangement of work environment elements and the occupational risk distribution pattern obtained as a result of operation of the genetic algorithm for the minimum allowable machine–machine distance of 20 m.

possible to "squeeze" the machines into an area guaranteeing that all workstations will be situated within the low occupational risk area. With the increasing value of the minimum allowable distance between machines, a "conflict" grows up between the criterion of low exposure to noise at workstations and the requirement that the machines are not situated too close to each other. For the minimum allowable distance between any two machines equaling 5 m, the effect of the penalty function becomes more significant. In the example simulation (Figure 5.11), the workstation w8 found itself in the medium occupational risk area, and workstation w6 on the edge of the area. When the minimum allowable distance between machines becomes comparable to room dimensions, the genetic algorithm is unable to reconcile requirements represented by the fitness function on one hand and the penalty function on the other. The penalty function dominates the fitness function and as a result of calculation, one obtains an arrangement of sources which fails to assure any high fitness value, as it can be seen from Figure 5.12. Despite the condition of keeping proper distance between machines has been met, workstation w7 is still situated in the high occupational risk area, and workstations w3, w5, and w8 are in the medium risk area.

Example 2—The effect of mutation probability on algorithm operation and result of optimization on example of noise sources and workstations arrangement optimization

Figures 5.13 through 5.15 illustrate the effect of the mutation occurrence probability on the genetic algorithm operation. For that specific simulation it was assumed

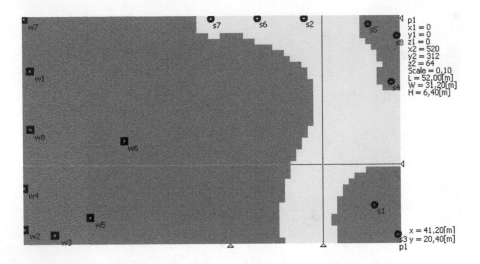

FIGURE 5.13 Arrangement of work environment elements and the occupational risk distribution pattern obtained as a result of operation of the genetic algorithm for the mutation probability of 0.001.

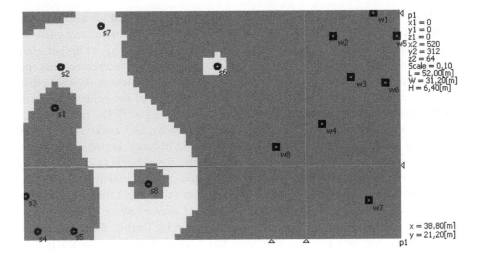

FIGURE 5.14 Arrangement of work environment elements and the occupational risk distribution pattern obtained as a result of operation of the genetic algorithm for the mutation probability of 0.01.

that the minimum distance between the work environment elements would be 4 m. In accordance with the adopted parameters and constraints imposed for the relative position of sound sources and workstations, the result of operation of the algorithm with very low mutation probability (0.001) is an arrangement of elements in which sound sources form a group located at the maximum possible distance from the group formed by workstations (Figure 5.13). For such locations, all workstations

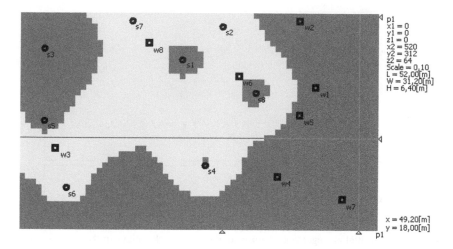

FIGURE 5.15 Arrangement of work environment elements and the occupational risk distribution pattern obtained as a result of operation of the genetic algorithm for the mutation probability of 0.2.

are located within a low occupational risk area. With increasing mutation probability, the ultimate result of operation of the algorithm becomes more and more distant from the optimum. Figure 5.14 shows a result of a simulation in which after termination of operation of the algorithm, source s6 is shifted toward workstations although that has a negative effect on the fitness function value. For the mutation probability of 0.2, mutation-induced random variations of optimization problem variables are large enough to make the algorithm divergent and finding the optimal solution becomes impracticable (Figure 5.15).

5.2.5 OPTIMIZATION OF NOISE SOURCES POSITIONS ORIENTED AT MINIMIZATION OF EXPOSURE TO NOISE—AN EXAMPLE

In the computer application, a model of a work room of dimensions 34.8 m × 13.9 m × 4.6 m was created. Sound absorption coefficients of room surfaces varied from 0.2 to 0.4 and the calculated average sound absorption coefficient equaled 0.3. Ten noise sources and seven workstations were placed in the room. In Figure 5.16, the initial arrangement of noise sources and workstations is shown. The A-weighted sound power levels of noise sources are presented in Table 5.1. Table 5.2 presents the A-weighted sound pressure levels at a workstation, calculated using Equations 5.16 and 5.18. The point noise sources were placed at a height of 1 m and sound pressure levels at workstations were calculated on a plane located at a height of 1.60 m. Risk areas for initial work room layout obtained from the calculation are shown in Figure 5.17.

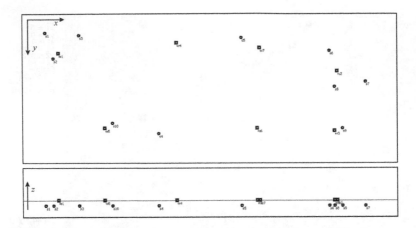

FIGURE 5.16   Initial arrangement of noise sources and workstations in the work room.

TABLE 5.1
Sound Power Level of Noise Sources

Noise source	s1	s2	s3	s4	s5	s6	s7	s8	s9	s10
$L_{W,A,j}$ (dB)	85	91	90	89	89	95	85	97	88	90

TABLE 5.2
Sound Pressure Levels on Workstations Calculated for Initial Work Room Layout

Workstation	w1	w2	w3	w4	w5	w6	w7
$L_{p,A,i}$ (dB)	84.10	85.11	82.67	80.03	82.15	80.19	80.89

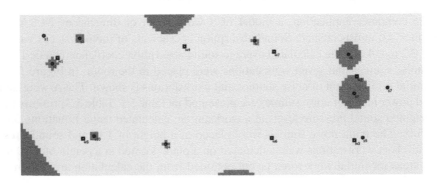

FIGURE 5.17   Risk areas distribution for initial arrangement of noise sources in the work room.

In the described example, the genetic algorithm was used for optimization of noise source location. Minimum distance between noise sources and room walls was set at 2 m, minimum distance between noise sources was set at 4 m, and minimum distance between noise source and workstation was set at 2 m. Simulations were made for different population size (L_{pop}), number of generations (L_{gnr}), crossover probability (P_{crs}), and mutation probability (P_{mut}). After each optimization, A-weighted sound pressure levels at workstation and fitness of best chromosome were calculated. The results of simulations are shown in Table 5.3 (S_n in the first column denotes the simulation number).

The simulation results given in Table 5.3 show that such optimization of noise source location carried out with the genetic algorithm, even with the small population size and number of generations, can significantly reduce workers' exposure to noise. If the highest sound pressure level existing at the workstation after the optimization process will be considered as a quality indicator of the given solution, the best results are achieved in simulations numbered 7 and 15. Additionally, because of the fact that in simulation number 5 sound pressure levels at two workstations, numbered 6 and 7, are over 81 dB, the solution achieved during simulation number 15 will be considered the better one. Arrangement of noise sources and the corresponding distribution of risk areas is presented in Figure 5.18.

According to presented results, population size equals 100 and number of generations equals 200, which make it sufficient to achieve high reduction of workers, exposure to noise. Further increase of the number of generations improves achieved

TABLE 5.3
Results of Simulations Research

S_n	L_{pop}	L_{gnr}	P_{crs}	P_{mut}	$L_{p,A,1}$ (dB)	$L_{p,A,2}$ (dB)	$L_{p,A,3}$ (dB)	$L_{p,A,4}$ (dB)	$L_{p,A,5}$ (dB)	$L_{p,A,6}$ (dB)	$L_{p,A,7}$ (dB)	Fitness
1	10	10	0.6	0.001	80.32	82.26	82.22	82.51	82.62	83.13	80.59	7.47
2	10	50	0.6	0.001	83.54	80.39	79.96	81.67	80.41	81.08	80.28	10.57
3	10	100	0.6	0.001	80.40	80.20	79.91	80.53	81.82	80.57	80.71	11.34
4	50	10	0.6	0.001	82.93	81.04	80.65	80.34	81.61	80.39	80.34	8.87
5	100	10	0.6	0.001	79.98	80.60	83.07	80.54	80.00	81.69	81.25	12.16
6	100	50	0.6	0.001	80.37	80.21	79.90	80.55	81.30	80.14	80.88	11.58
7	100	100	0.6	0.001	79.94	81.50	80.50	80.35	80.38	80.63	80.83	11.32
8	300	200	0.6	0.001	79.92	80.61	80.36	80.81	79.97	81.13	81.02	13.01
9	100	200	0.6	0.001	80.37	79.84	79.80	80.90	81.45	80.13	79.96	15.07
10	100	200	0.8	0.001	79.82	81.80	80.49	80.19	79.87	80.31	80.77	13.32
11	100	200	0.4	0.001	81.03	79.84	79.79	80.70	81.83	80.03	80.05	13.36
12	100	200	0.6	0.0005	79.91	80.79	81.22	80.20	79.98	80.70	80.61	13.15
13	100	200	0.6	0.005	79.96	80.41	79.99	80.82	80.35	80.51	85.26	12.16
14	100	200	0.6	0.01	79.96	80.93	80.25	80.72	80.00	80.48	81.45	13.91
15	100	500	0.6	0.001	79.92	80.69	80.33	80.51	79.96	81.20	80.35	13.29
16	200	500	0.6	0.005	81.31	79.84	79.79	80.54	81.20	79.97	79.99	16.79
17	500	1000	0.6	0.005	82.00	79.93	79.98	80.37	80.90	79.96	79.92	16.60

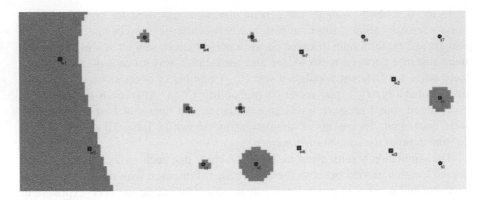

FIGURE 5.18 Arrangement of noise sources and distribution of risk zones obtained as a result of simulation No. 15.

results, but only to a small extent, and significantly increases calculation time. In this case, changes of the crossover probability have no significant influence on optimization results. An increase of mutation probability can cause degradation of the achieved results, as can be observed in simulation number 13. In this simulation, the best solution is achieved in the last genetic algorithms generation causing the noise exposure level to exceed the exposure limit value at workstation number 7.

The results of simulations numbered 16 and 17 showed that higher chromosome fitness, over 16, can be achieved for solutions which, according to above established quality indicator, are worse than solutions for which chromosome fitness amounts to about 13. It is caused by dominant influence of workstations at which sound pressure level is equal or below 80 dB on the fitness value. In order to overcome this disadvantage, the differences between the values of weighting factor μ, in case of those defined in Equation 5.6, should be larger.

5.3 SUMMARY

Results of the simulations described in this chapter show that genetic algorithms are an optimization tool which, when appropriately used, can be very useful for supporting actions concerning reduction of occupational exposure to noise. The most important factor for practical usage of genetic algorithms, especially in computer applications, is an appropriate definition of fitness function. This allows the reaching of the optimal solution with consideration of existing constraints on the real work environment. As the simulation results show, the genetic algorithm based on fitness function described in this chapter allows achievement of good results when applied to optimization of noise source location in order to reduce workers exposure to noise. The computer application presented in this chapter was elaborated on especially for this research, however, it made a fully functional program, which in limited range can be used for supporting action concerning the optimization of location of noise sources and workstations in the working environment for minimization of occupational risk resulting from noise.

REFERENCES

Asawarungsaengkul, K., and S. Nanthavanij. 2006. Design of optimal noise hazard control strategy with budget constraint. *Int J Occup Saf Ergon* 12(4):355–367. DOI:10.1080/10 803548.2006.11076695.

Aslan, F., and R. Paurobally. 2016. Modelling and simulation of active noise control in a small room. *J Vib Control* 24(3):607–618. DOI:10.1177/1077546316647572.

Bäck, T., D. B. Fogel, and Z. Michalewicz, eds. 2000. *Evolutionary Computation 1: Basic Algorithms and Operators*. New York: CRC Press, Taylor & Francis Group.

Baulac, M., J. Defrance, and P. Jean. 2007. Optimization of multiple edge barriers with genetic algorithms coupled with a Nelder-Mead local search. *J Sound Vibr* 300(1–2):71–87. DOI:10.1016/j.jsv.2006.07.030.

Baulac, M., J. Defrance, and P. Jean. 2008. Optimisation with genetic algorithm of the acoustic performance of T-shaped noise barriers with a reactive top surface. *Appl Acoust* 69(4):332–342. DOI:10.1016/j.apacoust.2006.11.002.

Berger, E. H., L. H. Royster, J. D. Royster, D. P. Dricsoll, and M. Layne, eds. 2003. *The Noise Manual*. Revised 5th edition. Fairfax: American Industrial Hygiene Association.

Blevins, M. G. 2016. Design and optimization of membrane-type acoustic metamaterials. *Architectural Engineering: Dissertations and Students Research*. 38. University of Nebraska, Lincoln, http://digitalcommons.unl.edu/archengdiss/38 (accessed May 29, 2020).

Chang, Y.-Ch., L.-J. Yeh, and M.-Ch. Chiu. 2005. Shape optimisation on constrained single-chamber muffler by using GA method and mathematical gradient method. *Int J Acoust Vib* 10(1):17–25.

Chang, Y.-Ch., M.-Ch. Chiu, and Ji-L. Xie, 2017. Noise elimination of reciprocating compressors using FEM, neural networks method, and the GA method. *Arch Acoust* 42(2):189–197. DOI:10.1515/aoa-2017-0021.

Chiu, M. Ch. 2010. Shape optimization of multi chamber mufflers with plug-inlet tube on a venting process by genetic algorithm. *Appl Acoust* 71(6):495–505.

Davis, D. 2009. Optimisation of noise control treatments for staged noise management programs using genetic algorithms. *Proceedings of ACOUSTICS 2009*, November 23–25, 2009, Adelaide, Australia.

Davis, L. ed. 1991. *Handbook of Genetic Algorithms*. New York: Van Nostrand Reinhold.

Directive 2003/10/EC of the European Parliament and of the Council of 6 February 2003 on the minimum health and safety requirements regarding the exposure of workers to the risks arising from physical agents (noise) (Seventeenth individual Directive within the meaning of Article 16(1) of Directive 89/391/EEC). OJ 42/38, 15.2.2003.

Engel, Z., D. Koradecka, D. Augustyńska, P. Kowalski, L. Morzynski, and J. Żera. 2010. Vibroacustic hazards. In *Handbook of Occupational Safety and Health*, ed. D. Koradecka, 153–189. Boca Raton: CRC Press, Taylor & Francis Group.

Goldberg, D. E. 1989. *Genetic Algorithms in Search, Optimization and Machine Learning*. Reading: Addison Wesley.

Grubeša, S., K. Jambrošić, and H. Domitrović. 2012. Noise barriers with varying cross-section optimized by genetic algorithm. *Appl Acoust* 73(11):1129–1137. DOI:10.1016/j.apacoust.2012.05.005.

Gundogu, Ö., M. Gokdag, and F. Yuksel. 2005. A traffic noise prediction method based on vehicle composition using genetic algorithms. *Appl Acoust* 66(7):799–809. DOI:10.1016/j.apacoust.2004.11.003.

Gwiazda, T. D. 2006. *Genetic Algorithms Reference, Volume I, Crossover for Single – Objective Numerical Optimization Problems*. Łomianki: Tomasz Gwiazda. www.tomaszgwiazda.com/Genetic_algorithms_reference_first_40_pages.pdf (accessed January 28, 2020).

Gwiazda, T. D. 2007. *Genetic Algorithms Reference, Volume II, Mutation Operator for Numerical Optimization Problems*. Łomianki. Tomasz Gwiazda.

Haupt, R. L., and S. E. Haupt. 1998. *Practical Genetic Algorithms*. New York: John Wiley & Sons. http://index-of.es/z0ro-Repository-3/Genetic-Algorithm/R.L.Haupt,%20 S.E.Haupt%20-%20Practical%20Genetic%20Algorithms.pdf (accessed January 28, 2020).

Holland, J. H. 1962. Outline for a biological theory of adaptive systems. *J ACM* 3:297–314. DOI:10.1145/321127.321128.

Holland, J. H. 1975. *Adaptation of Natural and Artificial Systems: An Introductory Analysis with Applications to Biology, Control, and Artificial Intelligence*. Ann Arbor: University of Michigan Press.

Lagarrigue, C., J.-P. Groby, O. Dazel, and V. Tourant. 2016. Design of metaporous super-cells by genetic algorithm for absorption optimization on a wide frequency band. *Appl Acoust* 102:49–54. DOI:10.1016/j.apacoust.2015.09.011.

Lan, T.-S., and M.-Ch. Chiu. 2008. Identification of noise sources in factory's sound field by using genetic algorithm. *Appl Acoust* 69(8):733–750. DOI:10.1016/j.apacoust.2007.02.007.

Le Courtois F., J. H. Thomas, F. Poisson, and J.-C. Pascal. 2016. Genetic optimisation of a plane array geometry for beamforming. Application to source localisation in a high speed train. *J Sound Vibr* 371:78–93. DOI:10.1016/j.jsv.2016.02.004.

Liu, H.-L. 2016. Acoustic partial discharge localization methodology in power transform-ers employing the quantum genetic algorithm. *Appl Acoust* 102:71–78. DOI:10.1016/j.apacoust.2015.08.011.

Makarewicz, G. 2007. Application of genetic algorithm in an active noise control system. *Arch Acoust* 32(4):839–849.

Morzyński, L. 2010a. Algorytmy genetyczne do minimalizacji ryzyka zawodowego związanego z ekspozycją na hałas – materiały informacyjne. [Genetic algorithms for minimization of occupational risk connected with exposure to noise—information materials]. www.ciop.pl/CIOPPortalWAR/file/46917/NA_M_2R18.pdf (accessed January 28, 2020).

Morzyński, L. 2010b. The use of genetic algorithm for limitation of occupational exposure to noise – Simulation research. *Proceedings of 15th International Conference on Noise Control, Noise Control'10*, June 6–9, 2010, Zamek Książ – Wałbrzych, Poland.

Morzyński, L. 2011. Wykorzystanie algorytmów genetycznych do ograniczania zawodowej ekspozycji na hałas. [Using genetic algorithms to limit occupational exposure to noise]. *Bezpieczeństwo Pracy – Nauka i Praktyka* 10(481):9–12.

NIOSH [National Institute of Occupational Safety and Health]. 1998. Criteria for a Recommended Standard. Occupational Noise Exposure. Revised Criteria. Cincinnati: NIOSH. www.cdc.gov/niosh/docs/98-126/pdfs/98-126.pdf (accessed February 3, 2020).

Peters, R. J., B. J. Smith, and M. Hollins. 2011. *Acoustics and Noise Control*. 3rd ed. Abington: Routledge, Taylor & Francis Group.

Razavi, H., E. Ramezanifar, and J. Bagherzadeh. 2014. An economic policy for noise control in industry using genetic algorithm. *Saf Sci* 65:79–85. DOI:10.1016/j.ssci.2013.12.010.

Sarafraz, H., Z. Sarafraz, M. Hodaei, and M. Sayeh. 2016. Minimizing vehicle noise passing the street bumps using Genetic Algorithm. *Appl Acoust* 106:87–92. DOI:10.1016/j.apacoust.2015.11.021.

Sivanandam, S. N., and S. N. Deepa. 2008. *Introduction to Genetic Algorithms*. Berlin: Springer.

Vér, I. L., and L. Beranek, eds. 2006. *Noise and Vibration Control Engineering, Principles and Applications*. 2nd ed. Hoboken: John Wiley & Sons.

Waly, S. M., and B. R. Sarker. 1998. Noise reduction using nonlinear optimization modeling. *Comput Ind Eng* 35(1–2):327–330. DOI:10.1016/S0360-8352(98)00086-2.

Yeniay, Ö. 2005. Penalty function methods for constrained optimization with genetic algorithm. *Mathematical and Computational Applications* 10(1):45–56. DOI:10.3390/mca10010045.

Zangeneh-Nejad, F., and R. Fleury. 2019. Active times for acoustic metamaterials. *Reviews in Physics* 4:e100031. DOI:10.1016/j.revip.2019.100031.

6 A Multi-Index Method for Acoustic Quality Assessment of Classrooms

Jan Radosz

CONTENTS

6.1 INTRODUCTION

The acoustic quality of a room intended mainly for verbal communication (a classroom, lecture hall, speech therapy room, etc.) can be defined as a parameter used to describe its acoustic properties connected with subjective impression experienced in scope of, among other things, intelligibility of speech, level of noise disrupting verbal messages, or the speaker's speech effort. The acoustic quality of a classroom is also a carrier of information whether the space meets conditions guaranteeing, in particular:

- adequate speech intelligibility,
- low background noise level in the course of classes,

- no need to speak with raised voice, and
- teaching and learning comfort.

The acoustic quality of classrooms is affected by many factors such as room volume (cubature), equipment, or noise penetrating from outside [Kotus et al. 2010; Leśna and Skrodzka 2010; Mikulski and Radosz 2010]. Inadequate acoustic quality is sometimes a result of, for instance, wrong architectural solutions resulting in, for example, excessively long reverberation time and the related higher background noise level.

One method of reverberation reduction and noise suppression in classrooms and improvement of conditions for verbal communication consists in proper shaping of acoustic properties of these spaces [Mikulski and Radosz 2011].

Although the acoustics of classrooms is an issue discussed extensively in scientific journals for many years now, research initiatives in that area are still undertaken in many countries (including Greece, Italy, England, China, and Brazil) which evidences continuous relevance of the topic. The studies concern mainly the analysis of acoustic parameters in existing premises and technical solutions aimed at their improvement.

Generally speaking, assessment of acoustics of a room may be based on objective and/or subjective criteria [Bradley 1986; Sato and Bradley 2008; Rudno-Rudzińska and Czajkowska 2010]. The objective parameters are well-defined, assume specific numerical values, and are determined by way of standardized measuring methods. Subjective parameters are those which qualitatively express subjective assessment of the acoustic properties of a room and are represented by epithets representing the verbal equivalent of a parameter characterizing the corresponding acoustic property.

Among objective parameters, one might include the reverberation time (T_{60}, T_{20}, T_{mf}), the speech transmission index STI, the clarity index C_{50}, or the sound pressure level L_p in the room.

More subjective parameters, such as the liveness, the intimacy, or the clarity of sound were defined for rooms in which live music is performed and listened to (concert halls and opera theatres). For that reason, these parameters are not used in acoustic assessment of classrooms where verbal communication is definitely the key issue.

Room acoustics is the subject of numerous regulatory documents (standards, recommendations, building regulations) applicable in different countries and concerning criteria to be used in assessment of acoustical parameters of rooms. In the European Union countries, examples of such national regulations are the Building Bulletin 93 (United Kingdom), DIN 18041:2004-05 standard (Germany), or SFS 5907:2006 standard (Finland). Normative documents concerning building acoustics pass over many acoustic parameters essential from the point of view of the room acoustic quality and in the majority of cases, requirements set out in the standards are limited to required values of the reverberation time, acceptable sound pressure levels in rooms, and the required insulation performance of partitions in a building which are far insufficient to ensure uniqueness of acoustic assessment results.

One way to obtain a unique quantitative assessment of the acoustic quality of a room or other utility space is to apply the multi-index method which consists in determination of value of a single-number global index based on a number of partial indices. The multi-index method can be employed as an evaluation tool in many situations typical for vibro-acoustic studies, such as the assessment of acoustic quality of machines, long narrow rooms, enclosures open on both ends (underground stations), and industrial structures. The method is the subject of continuous elaboration and refinement in both Polish and foreign research centers. Its key merit consists in comprehensiveness of approach to the acoustic quality issue, as well as uniqueness and comparability of the assessment result. This chapter presents results of studies on application of the multi-index method to assessment of acoustic quality of classrooms [Radosz 2013].

6.2 REQUIREMENTS AND RECOMMENDATIONS CONCERNING OBJECTIVE ACOUSTIC PARAMETERS OF CLASSROOMS

6.2.1 ACCEPTABLE SOUND PRESSURE LEVEL VALUES

Acceptable values of the sound pressure level in classrooms are scattered throughout different regulations, directives, standards, and guidelines concerning different types of rooms depending on their types and functions.

Guidelines of the World Health Organization (WHO) determine the recommended sound pressure level value—expressed in terms of the A-weighted equivalent sound pressure level of noise penetrating into a room—as equaling 35 dB. The same value is recommended in the United States in ANSI S12.60-2002. In the majority of European Union countries, the acceptable sound pressure levels in rooms (expressed in terms of the A-weighted equivalent sound pressure level of noise penetrating into a room) are included in the range 35–40 dB, as provided, for instance, in PN-B-02151-02:1987.

6.2.2 REVERBERATION TIME AND SPEECH INTELLIGIBILITY

Acceptable reverberation time values adopted in various countries are included in the range 0.6–0.8 s, depending on function and cubature of the room and the frequency bands to which the criteria apply (Table 6.1).

Additionally, it is recommended that in rooms intended for people with hearing loss and/or other verbal communication issues, the maximum reverberation time should not exceed 0.4 s.

The majority of requirements or recommendations concerning acoustics of classrooms pertains only to the reverberation time, sound-insulating properties of partitions, and acceptable values of the sound pressure level in rooms. Only a few requirements concern the intelligibility of speech. The latter can be quantitatively characterized with the use of such objective parameters as the speech transmission index STI or the clarity index C_{50}. In British guidelines BB93, the minimum value of STI acceptable for classrooms is 0.6. Finnish guidelines set out in SFS 5907 standard determine $STI = 0.8$ as the minimum value of the index. In Polish standard PN-B-02151-4:2015, the required minimum value of the index is $STI = 0.6$.

TABLE 6.1

Acceptable Values of the Reverberation Time and the Speech Transmission Index *STI* in Classrooms in Selected Countries

Country	Reference Document	Required Reverberation Time (s)	Required STI Value
Denmark	Standard BR 2010	0.6 (125–4000 Hz)	—
Finland	Standard SFS 5907:en	0.6–0.8 (for 250–4000 Hz) by 50% higher for 125 Hz	≥0.8 (for class A and B rooms defined in the standard)
France	Decree CCH Arreté du 25 avril 2003	$0.4 \leq T_{mf} \leq 0.8$ (for $V < 250$ m³) $0.6 \leq T_{mf} \leq 1.2$ (for $V < 250$ m³)	—
Germany	Standard DIN 18041	$T_{opt} = (0.32 \log_{10}\{V\} - 0.17)$ s	—
Spain	Guidelines CTE DB-HR	$T_{mf} \leq 0.5$ (for $V < 350$ m³)	—
USA	Standard ANSI S.12.60	$T_{mf} < 0.6$ (for $V < 283$ m³) $T_{mf} < 0.7$ (for $V > 283$ m³ and $V \leq 566$ m³)	—
United Kingdom	Guidelines Building regulations BB93	$T_{mf} < 0.6$ for primary schools $T_{mf} < 0.8$ for secondary schools	≥0.6
Poland	Standard PN-B-02151-4:2015	≤0.6	≥0.60

V—room cubature (m³); T_{opt}—optimum reverberation time value (s).

For the clarity index C_{50}, Bradley [1986] recommends the value of 1 dB for the frequency band centered at 1 kHz as the minimum for rooms in which verbal messages are to be conveyed.

6.3   THE MULTI-INDEX METHOD

The method for acoustic quality assessment of rooms presented in this chapter consists in determining a number of partial indexes and using them to calculate a global index which determines uniquely the acoustic quality of the assessed room. The assumptions for the proposed assessment methodology were adopted on the grounds of results of experimental studies [Mikulski and Radosz 2011; Mikulski 2012] and other index methods [Pleban 1999, 2010, 2011; Piechowicz 2004; Engel et al. 2007; Kosała 2008, 2011, 2012; Carvalho and Silva 2010]. To work out the assumptions, results of measurements of acoustic parameters in more than 100 classrooms were taken into account as well as analyses of data published in the literature concerning factors affecting intelligibility of verbal messages, background noise level, teacher's speech effort, and comfort of both teaching and learning. As a result, it is assumed that to carry out a quantitative acoustic quality assessment of a room with the use of the multi-index method it is necessary to measure:

- the impulse response of the room and use it to determine the reverberation time T_{20}, the speech transmission index STI, the clarity index C_{50}, and the relative sound strength G_{rel},
- the A-weighted sound pressure level to determine the speech effort,
- the signal-to-noise ratio (SNR) in the course of classes, and
- the A-weighted sound pressure level of noise background in the room from all noise sources combined.

6.4   THE GLOBAL ACOUSTIC QUALITY INDEX OF CLASSROOMS

The global acoustic quality index QI_{G} is determined based on partial indexes with their individual weights taken into account [Radosz 2013] and calculated according to the formula

$$QI_{G} = \frac{\sum_{i=1}^{n} QI_i \eta_i}{\sum_{i=1}^{n} \eta_i} \tag{6.1}$$

where:
   QI_i: i-th partial index
   η_i: weight of i-th partial index
   n: number of partial indexes

To determine the global acoustic quality index of a classroom, six partial indexes are proposed by the present author to be taken into account. As was already mentioned above, assumptions adopted to identify and define individual partial indexes are based on analysis of data published in the literature of the subject and on results of the author's own research work, with the basic premise being to obtain a comprehensive picture of the acoustic quality of assessed rooms. After a detailed analysis of available data it is assumed that the following factors will be taken into account in assessing acoustic quality of a room:

- reverberation conditions prevailing in the room (the reverberation index),
- conditions of verbal communication (the speech intelligibility index),
- external and internal interference (the background noise index),
- speaker's (teacher's, lecturer's) speech effort (the speech effort index),
- the difference between the speech signal level and the background noise level in the course of conducting classes (the signal-to-noise ratio index), and
- differences in sound perception depending on the location occupied by the listener in the room (the sound strength distribution index).

With the above-listed factors taken into account, Equation 6.1 takes the form

$$QI_G = \frac{QI_{RT}\eta_{RT} + QI_{SI}\eta_{SI} + QI_{SE}\eta_{SE} + QI_{SD}\eta_{SD} + QI_{BN}\eta_{BN} + QI_{SNR}\eta_{SNR}}{\eta_{RT} + \eta_{SI} + \eta_{SE} + \eta_{SD} + \eta_{BN} + \eta_{SNR}} \quad (6.2)$$

where:
   QI_{RT}: the reverberation index
   QI_{SI}: the speech intelligibility index
   QI_{SE}: the speech effort index
   QI_{SD}: the sound strength distribution index
   QI_{BN}: the background noise index
   QI_{SNR}: the signal-to-noise ratio index
   η_{RT}: the reverberation index weight
   η_{SI}: the speech intelligibility index weight
   η_{SE}: the speech effort index weight
   η_{SD}: the sound strength distribution index weight
   η_{BN}: the background noise index weight
   η_{SNR}: the signal-to-noise ratio index weight

By way of analogy to other applications of the multi-index method and other dimensionless indexes (e.g. *STI*) it is assumed that dimensionless partial indexes and the global index will assume values from the range 0–1. The better the acoustic quality of a room, the higher will be the values of partial indexes and ultimately, of the global index. To ensure uniqueness of the proposed room acoustics quality assessment method, an evaluation scale is adopted in which specific ranges of the global quality index QI_G values are ascribed to corresponding descriptive interpretations of acoustic quality (Figure 6.1). The categorization is based on analysis of both actual acoustic parameters characterizing the examined rooms and recommendations and requirements set out in standards for rooms of that type.

Weights of partial indexes assume values from the range 0–1. The higher the weight value, the more the corresponding partial index contributes to the acoustic quality assessment result. The effect of weights of partial indexes on acoustic quality is depicted in Figure 6.2.

The weights proposed to be assigned to individual partial indexes are presented in Table 6.2. Values of the weights do not follow from any strict relationships but

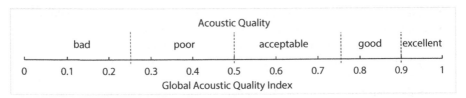

FIGURE 6.1 Values of the global index QI_G ascribed to the room acoustic quality interpretation scale. (From Radosz, J., *Arch. Acoust.*, 38, 159–168, 2013.)

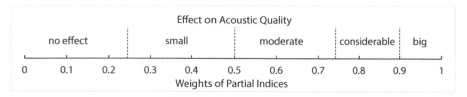

FIGURE 6.2   The effect of weights of partial indexes on acoustic quality of classrooms.

TABLE 6.2
Weights of Partial Indexes

η_{RT} Reverberation Index Weight	η_{SI} Speech Intelligibility Index Weight	η_{SE} Speech Effort Index Weight	η_{SD} Sound Distribution Index Weight	η_{BN} Background Noise Index Weight	η_{SNR} Signal-to-Noise Index Weight
0.8	1	0.3	0.5	1	0.5

were adopted as a result of analysis of factors showing the most important effect on acoustic quality of rooms (including correlation and statistical significance of comparable acoustic parameters) and based on results of experimental studies carried out in selected rooms. The issue of determination of values of individual weights will be discussed in detail in the next section.

6.5   PARTIAL INDEXES

6.5.1   THE REVERBERATION INDEX

The reverberation time, in view of its strong correlation with aural sensations, is one of the most important criteria in any room acoustics assessment method. As a rule, the parameter is determined in a number of octave frequency bands. Studies carried out by Sato and Bradley [2008] prove that difficulties with intelligibility of speech should be linked with, more than anything else, values of the reverberation time in the frequency range of 1–4 kHz. The same studies also indicate a strong correlation of the reverberation time in the octave band with mid-band frequency of 2 kHz with results from the subjective speech intelligibility test. Results of tests presented in this chapter confirm such observations. In view of the above, to determine the reverberation index value, the reverberation time measured in the octave band with the mid-band frequency of 2 kHz is adopted.

The relationship between the reverberation time $RT_{2\,kHz}$ and the reverberation index QI_{RT} was determined empirically based on results of research carried out both in Poland and abroad as well as criteria and requirements applicable to rooms of that type. For the purpose of determination of the relationship, the acoustic absorption of pupils was assumed to be 0.41 m² per occupant in the 2 kHz octave frequency band. The reverberation index QI_{RT} assumes the value of 1 for the reverberation time $RT_{2\,kHz}$ from the range 0.45–0.55 s which is the optimum reverberation time value for classrooms with volume not exceeding 200 m^3, coinciding with the criteria and requirements defined in applicable standards (cf. Section 6.2).

The relationship between the reverberation time $RT_{2\,kHz}$ and the reverberation index QI_{RT} can be described with the use of the formula:

$$QI_{RT} = -0.48\{RT_{2\,kHz}\}^4 + 2.55\{RT_{2\,kHz}\}^3 - 4.77\{RT_{2\,kHz}\}^2 + 3.13\{RT_{2\,kHz}\} + 0.34$$

$$(6.3)$$

where $RT_{2\,kHz}$ (s) is the reverberation time in the octave band with the mid-band frequency of 2 kHz.

Although the reverberation time is one of the most important parameters of room acoustics, it follows from the published data that the quantity is not correlated with subjective speech intelligibility as strongly as other acoustic parameters (it shows lower values of the correlation coefficient). The fact is confirmed also by the present author's own studies. Therefore, assuming that the index cannot assume the value of 1 (the highest possible value for any weight) and taking into account the correlation between results of subjective tests and the reverberation time, the reverberation index weight value η_{RT} is assumed to be 0.8.

6.5.2 THE SPEECH INTELLIGIBILITY INDEX

Intelligibility of speech is a very important issue in the process of verbal communication. To evaluate the speech intelligibility objectively, the speech transmission index *STI* is used which is a quantity strongly correlated with subjective impressions in the area of comprehensibility of oral messages [EN 60268-16:2011]. Values of the index may vary from 0 to 1, where 1 corresponds to excellent intelligibility [EN ISO 9921:2003].

Another parameter which can also be used for objective speech intelligibility assessment is the clarity index C_{50} defined as the ratio of signal energy reaching a listener during the first 50 ms to the total signal energy (value of the time interval corresponds to the time constant of human ear) [Bradley and Bistafa 2002]. The clarity index C_{50} is an important criterion of acoustic assessment of classrooms as it characterizes perception of sounds following quickly one after another in succession. As with the reverberation time, the value of the index is referred to in terms of individual octave frequency bands.

In view of strong correlation with subjective speech intelligibility, the octave band with the mid-band frequency of 1 kHz is adopted to determine the speech intelligibility index value. To determine the value of the speech intelligibility index QI_{SI} it is necessary to determine values of the auxiliary clarity index C_{aux}. The index is determined based on the value of the clarity index $C_{50,1\,kHz}$, under the assumption that for $C_{50,1\,kHz} \geq 4$ dB, the value of the auxiliary clarity index C_{aux} will be 1. To calculate C_{aux} for lower values of the clarity index, Bradley and Bistafa [2002] proposed the formula

$$C_{aux} = -0.00616\{C_{50,1\,kHz}\}^2 + 0.0615\{C_{50,1\,kHz}\} + 0.85 \qquad (6.4)$$

where $C_{50, 1 \, kHz}$ (dB) is the clarity index in the octave band with the mid-band frequency of 1 kHz.

Based on results of studies published in the literature of the subject it is assumed that the speech intelligibility index value depends on the speech transmission index *STI* and the clarity index $C_{50, 1 \, kHz}$. Taking into account the correlation between subjective intelligibility tests and the parameters describing the speech intelligibility quantitatively, it is assumed that QI_{SI} is determined from the following formula (with the rule of weighting of Section 6.3 and slightly less weight of the clarity index following from studies on speech intelligibility taken into account):

$$QI_{SI} = \frac{1}{1.8}\left(STI + 0.8 \, C_{aux}\right) \tag{6.5}$$

where:
   STI: the speech transmission index
   C_{aux}: the auxiliary clarity index

In view of what classrooms are intended for, the speech intelligibility is of primary importance. Since the objective acoustic parameters characterizing intelligibility of speech show the best correlation with subjective impressions and speech intelligibility tests, it is assumed that the value of the speech intelligibility index weight η_{SI} is 1.

6.5.3 THE SPEECH EFFORT INDEX

The sound pressure level is one of objective parameters characterizing the speech effort. According to EN ISO 9921:2003, the normal speech effort corresponds to the A-weighted sound pressure level of 60 dB measured at the distance of 1 m from speaker's mouth (Table 6.3).

With the use of the above-quoted data, a relationship is proposed to determine the speech effort index QI_{SE} based on the A-weighted sound pressure level of voice at the distance of 1 m from speaker's mouth. The relationship can be described

TABLE 6.3

The Speech Effort of Male Speaker and the Related A-Weighted Sound Pressure Levels at the Distance of 1 m from Speaker's Mouth

Speech Effort	A-Weighted Sound Pressure Level
Very loud speech	78 dB
Loud speech	72 dB
Raised voice	66 dB
Normal speech	60 dB

Source: ISO 9921:2003. Ergonomics—Assessment of speech communication.

by the following formula (under the assumption that for the teacher's voice level $L_{p,A,V,1\,m} = 60$ dB and lower, the speech effort index will be the unity, $QI_{SE} = 1$):

$$QI_{SE} = 3.46 - 0.041\{L_{p,A,V,1\,m}\} \tag{6.6}$$

where $L_{p,A,V,1\,m}$ (dB) is the A-weighted teacher's voice sound pressure level at the distance of 1 m.

Excessive speech effort lasting for a longer period of time can be the cause of chronic vocal organ diseases [Hodgson et al. 1999; Bronder 2003; Radosz 2012]. The speech effort depends largely on speakers themselves and the degree of control they maintain of their voice. Data available in the literature and results of the present author's own research do not provide any unambiguous evidence in support of possible negative effect of acoustic properties of rooms on the speech effort (low values of the correlation coefficient). However, for the same persons subject to tests in different reverberation conditions, different speech effort-related parameters can be obtained. As it is impossible to determine unambiguously any sufficiently accurate relationship between acoustic parameters of room and the speech effort of speaker, it is assumed that the effect of the corresponding partial index will be relatively small, and the value of the speech effort index weight η_{SE} is 0.3.

6.5.4 THE SOUND STRENGTH DISTRIBUTION INDEX

Another criterion important in assessment of acoustic quality of classrooms is the sound pressure level distribution in the room volume. The more uniform is the sound pressure level distribution, the better is the room acoustic quality. The parameter adopted as a base for evaluation of the sound pressure level distribution in a room is the relative sound strength in view of the possibility of determining it from the room impulse response function. The parameter is calculated typically for octave frequency bands [ISO 3382-1]. To evaluate the sound strength distribution in a room, the difference between extreme values of the relative sound strength $\Delta G_{rel,mbf}$ in octave frequency band with given mid-band frequency (mbf) was used.

To determine the values of the sound strength distribution index in given octave frequency band $QI_{SD,mbf}$, a functional relationship following from experimental studies is proposed on the assumption that if the sound strength distribution in a room is perfectly even, the sound strength distribution index assumes the value of 1. The relationship can be expressed with the formula

$$QI_{SD,mbf} = 1 - 0.08\{\Delta G_{rel,mbf}\} \tag{6.7}$$

where $\Delta G_{rel,mbf}$ (dB) is the difference between extreme values of the relative sound strength ΔG_{rel} for the octave frequency band with given mid-band frequency (mbf).

To determine values of the sound strength distribution index QI_{SD}, it is necessary to consider the frequency bands most important from the point of view of verbal communication and take into account their weights determined based on correlation of acoustic parameters with subjective impressions and objective speech intelligibility tests. With the data available in the literature taken into account it is assumed that the frequency bands of 1 kHz, 2 kHz, and 4 kHz will be considered (with the most important effect of 2 kHz and 4 kHz bands and slightly less of the 1 kHz band) in view of their fundamental significance for verbal communication, and the sound strength distribution index QI_{SD} will be expressed by means of the formula (which, like in case of the speech intelligibility index, uses the principle of weighing indexes introduced in Section 6.3)

$$QI_{SD} = \frac{1}{2.7}\left(0.8\ QI_{SD,1\,kHz} + QI_{SD,2\,kHz} + 0.9\ QI_{SD,4\,kHz}\right) \qquad (6.8)$$

where $QI_{SD,1\,kHz\,(2\,kHz,\,4\,kHz)}$ is the sound strength distribution index for the octave band with the mid-band frequency of 1 kHz (2 kHz, 4 kHz).

In a room, the sound strength distribution is a parameter reflecting the effect of speech intelligibility deterioration with increasing distance from the speaker to the listener. However, in view of cubature values typical for classrooms (155–200 m³), the parameter is less significant in the global acoustic quality assessment compared to, for example, the reverberation index or the speech intelligibility index. The value for the sound strength distribution index weight η_{SD} is therefore assumed to be 0.5 based on experiments.

6.5.5   THE BACKGROUND NOISE INDEX

With the requirements concerning the noise background in a room from all noise sources taken into account (cf. Section 6.2), a formula is proposed to be used to determine the index representing the noise background in a room denoted QI_{BN} and called in short the background noise index. The relationship between the A-weighted sound pressure level of noise background in a room from all noise sources combined $L_{p,A,BN}$ and the background noise index QI_{BN} can be described with the use of the following formula expressing the assumption that for $L_{p,A,BN}$ $\leq$ 30 dB the index assumes the value $QI_{BN} = 1$, and for $L_{p,A,BN}$ > 60 dB its value becomes zero:

$$QI_{BN} = 0.0008\{L_{p,A,BN}\}^2 - 0.1083\{L_{p,A,BN}\} + 3.5 \qquad (6.9)$$

where $L_{p,A,BN}$ (dB) is the A-weighted sound pressure level of noise background in the room in question from all noise sources combined.

Any noise existing in a room in the course of the speech comprehension process affects perception of content included in verbal message [Koszarny 1992; Augustyńska et al. 2010]. What is more, admissible values of the A-weighted sound pressure level of noise background in rooms are subject to standardization (cf. Section 6.2.1). In view of the above, the weight of the index representing the noise background in a room is assumed to be the unity, $\eta_{BN} = 1$.

6.5.6 THE SIGNAL-TO-NOISE RATIO INDEX

The parameter known as the signal-to-noise ratio and denoted *SNR* is a quantity representing the distance between the speech signal level and the background noise level expressed in decibels. To measure the parameter, any sound meter/analyzer can be used provided it has the function of recording the sound pressure level waveforms. The measurement consists in recording the sound pressure level values with 200-ms sampling in the course of classes (Figure 6.3). Next, based on histograms of the A-weighted sound pressure level for the speech signal and the background noise level, the actual differences between the two quantities are determined. The measurement point in a given room is determined based on the criterion of highest difference between extreme values of the relative sound strength G_{rel}.

The *SNR* value optimal from the point of view of ensuring correct reception of the content of verbal message should not be lower than 15 dB. In view of the above, the following formula is adopted to determine values of the signal-to-noise ratio index QI_{SNR} (on the assumption that for $SNR = 15$ dB and higher values, the index assumes the value $QI_{SNR} = 1$):

$$QI_{SNR} = 0.058 \exp\left(0.18\{SNR\} + 0.14\right) \tag{6.10}$$

where $\{SNR\}$ (dB) is the value of the signal-to-noise ratio in the course of classes.

Value of the signal-to-noise ratio index QI_{SNR} depends partly on speakers (teachers) themselves and the degree to which they keep their voice under control. Based on the analysis of results obtained for the parameter *SNR* in examined rooms and with the data available in the literature of the subject taken into account, the value of the signal-to-noise ratio index weight η_{SNR} is assumed to be 0.5.

Where the present acoustic quality assessment method is to be applied in rooms in which it is impracticable to measure the signal-to-noise ratio or the assessment results serve a purpose of orienteering nature only (e.g., in the room design stage), it is recommended to assume that the value of the signal-to-noise ratio index $QI_{SNR} = 1$.

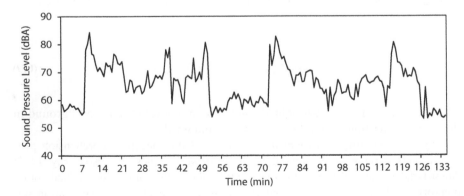

FIGURE 6.3 An example of the A-weighted sound pressure level changes in a room (school classroom).

6.6 EXAMPLES OF ACOUSTIC QUALITY ASSESSMENT FOR SELECTED ROOMS

For the purpose of a trial assessment, nine classrooms were selected as a result of application of the criterion of obtaining the widest possible range of acoustic parameters identified above as those necessary to determine partial indexes. The rooms (listed in Table 6.4) lacked any acoustic treatment. Results of measurements of acoustic parameters characterizing the rooms are summarized in Table 6.5.

Based on the obtained results of measurements of the above-listed acoustic parameters, partial indexes for individual rooms were determined (shown below in Figures 6.4 through 6.9). Next, with their weights taken into account, Equation 6.2 was used to calculate the global acoustic quality index for each of the rooms, with the result shown in Figure 6.10.

Results of the study revealed a diversity in both values of individual partial indexes and the resultant global acoustic quality index characterizing the tested classrooms. Values of the reverberation index in the analyzed rooms varied within the range from 0.50 to 0.97. The diversity was mainly the result of differences in equipment (furniture, chair types, carpets, flooring materials, cork boards, etc.). Values of the speech intelligibility index ranged from 0.60 to 0.76. The observed dispersion of values of the indicators across the rooms was not too large which can be attributed mainly to their similar cubic capacity. The obtained values of the speech effort index varied from 0.63 to 0.75—in each of the rooms, teachers found it necessary to at least raise their voice. As the examined rooms had virtually the same volume, values of the sound strength distribution index did not show any significant dispersion between the spaces and were included in the range from 0.74 to 0.87. An exception in the overall diversification of the indexes was the background noise index as in none of the examined rooms, acceptable values of the A-weighted equivalent sound pressure level of background noise from all the sources combined were exceeded.

TABLE 6.4
The Classrooms Selected for Trial Assessment

Classroom Type	Room ID	Cubature (m³)	Number of Pupils
Primary school	A	160	24
	B	160	22
	C	157	28
	D	157	32
	E	158	30
	F	158	32
	G	157	34
	H	157	34
	I	157	38

Source: From Radosz, J., *Arch. Acoust.*, 38, 159–168, 2013.

TABLE 6.5
Results of Measurements of Acoustic Parameters in Tested Classrooms

Room ID	$T_{2\,kHz}$ (s)	STI	$C_{50,1\,kHz}$ (dB)	$L_{p,A,V,1\,m}$ (dB)	$\Delta G_{rel,1\,kHz}$ (dB)	$\Delta G_{rel,2\,kHz}$ (dB)	$\Delta G_{rel,4\,kHz}$ (dB)	$L_{p,A,BN}$ (dB)	SNR (dB)
A	1.22	0.54	−2.3	66.0	2.8	1.2	1.8	23.9	11.1
B	1.00	0.59	−1.0	63.8	2.2	1.3	1.4	24.0	13.5
C	1.08	0.56	−1.5	62.8	3.3	2.4	2.0	25.9	11.2
D	1.18	0.63	−3.1	69.1	2.5	2.2	1.6	25.7	16.3
E	1.46	0.51	−3.7	68.3	1.4	2.1	2.6	24.6	7.7
F	1.12	0.56	−1.8	62.4	1.9	2.0	2.8	23.6	12.4
G	0.65	0.65	0.8	60.6	3.5	2.9	3.6	27.2	13.2
H	1.18	0.54	−3.3	65.8	3.1	2.5	2.3	27.8	12.4
I	1.14	0.64	−2.4	61.7	3.0	2.2	2.1	26.9	12.1

$T_{2\,kHz}$ (s): reverberation time in the band with the mid-band frequency of 2 kHz.

STI: speech transmission index.

$C_{50,1\,kHz}$ (dB): clarity index in the frequency band with the mid-band frequency of 1 kHz.

$L_{p,A,V,1\,m}$ (dB): A-weighted teacher's voice sound pressure level at the distance of 1 m.

$\Delta G_{rel,mbf}$ (dB): difference between extreme values of the relative sound strength in the octave frequency band with given mid-band frequency (mbf).

$L_{p,A,BN}$ (dB): A-weighted sound pressure level of noise background from all noise sources combined.

SNR (dB): signal-to-noise ratio.

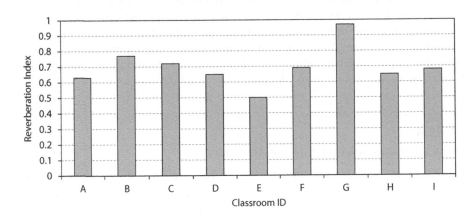

FIGURE 6.4   Values of the reverberation index for the tested classrooms.

Values of the global acoustic quality index in the examined rooms varied within the range 0.65–0.87. Therefore, none of the rooms met the conditions enabling it to be categorized as a room of excellent acoustic quality (Figure 6.10). On the other hand, none of the examined rooms was subject to acoustic treatment. It can be concluded from the above that in rooms with standard equipment and standard finishing of the building structure (not adapted acoustically to functions they serve) it is impossible to achieve the excellent acoustic quality rating.

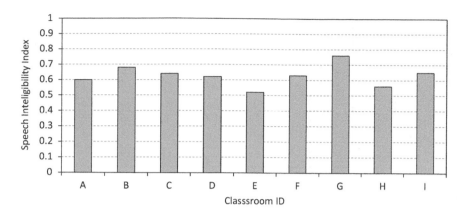

FIGURE 6.5   Values of the speech intelligibility index for the tested classrooms.

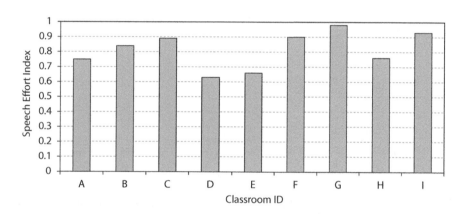

FIGURE 6.6   Values of the speech effort index for the tested classrooms.

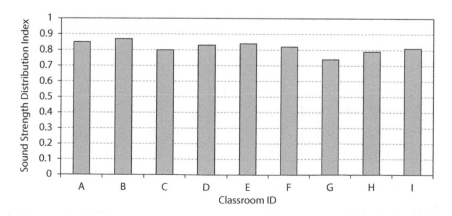

FIGURE 6.7   Values of the sound strength distribution index for the tested classrooms.

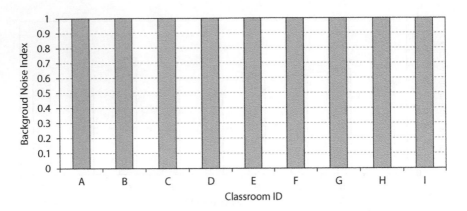

FIGURE 6.8 Values of the background noise index for the tested classrooms.

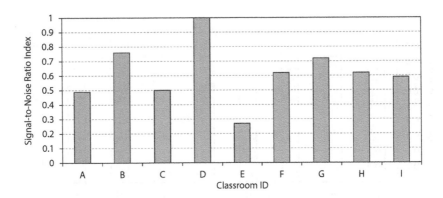

FIGURE 6.9 Values of the signal-to-noise ratio index for the tested classrooms.

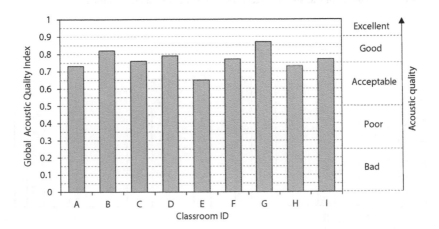

FIGURE 6.10 Values of the global acoustic quality index calculated for the tested classrooms.

6.7   SUMMARY

The above-presented method of assessment of the acoustic quality of classrooms offers new opportunities in the area of both assessment and improvement of acoustics in premises used for educational purposes. The assumptions adopted when developing the method ensure its objectivity (the method is based on objective acoustic parameters) and uniqueness (definite values of the global index are assigned to corresponding judgmental descriptions of the acoustic quality). The multi-index method for assessment of the acoustic quality of classrooms is a validated tool which can be used straight away in the room design and assessment process and in undertakings concerning technical prophylaxis. The tool offers architects, designers, and occupational safety and health officers the possibility to plan for comfortable acoustic conditions characterized with optimum acoustic parameters both in the stage of design and erection of new buildings used for educational purposes as well as in case of extension, modernization, and renovation of existing classrooms.

REFERENCES

ANSI S12.60-2002. American National Standard Acoustical Performance Criteria, Design Requirements, and Guidelines for Schools.

Augustyńska, D., A. Kaczmarska, W. Mikulski, and J. Radosz. 2010. Assessment of teachers' exposure to noise in selected primary schools. *Arch Acoust* 35(4):521–542. DOI:10.2478/v10168-010-0040-2.

Bradley, J. S. 1986. Predictors of speech intelligibility in rooms. *J Acoust Soc Am* 80(3):837–845. DOI:10.1121/1.393907.

Bradley, J. S., and S. R. Bistafa. 2002. Relating speech intelligibility to useful-to-detrimental sound ratios. *J Acoust Soc Am* 112(1):27–29. DOI:10.1121/1.1481508.

Bronder, A. 2003. Badanie przyczyn zaburzeń narządu głosu nauczycieli i opracowanie zasad profilaktyki. [A study on causes of vocal organ disorders in teachers and a draft of the principles of prevention]. PhD thesis, Instytut Medycyny Pracy i Zdrowia Środowiskowego w Sosnowcu, Sosnowiec.

Building Bulletin 93. 2015. Acoustic design of schools: Performance standards. Manchester: Department for Education. Education Funding Agency.

Carvalho, A., and P. Silva. 2010. Sound, noise and speech at the 9000-seat Holy Trinity Church in Fatima, Portugal. *Arch Acoust* 35(2):145–156. DOI:10.2478/v10168-010-0013-5.

DIN 18041:2004-05. 2004. Hörsamkeit in kleinen bis mittelgroßen Räumen. [German standard: Acoustic quality in small to medium-sized rooms]. Berlin, Germany: Deutsches Institut für Normung [German Institute for Standardization].

EN 60268-16:2011. 2011. Sound system equipment—Part 16: Objective rating of speech intelligibility by speech transmission index. Brussels, Belgium: European Committee for Standardization.

Engel, Z., J. Engel, K. Kosała, and J. Sadowski. 2007. Podstawy akustyki obiektów sakralnych. [Foundations of acoustics of sacral objects]. Cracow-Radom: Wydawnictwo Instytutu Technologii Eksploatacji – PIB.

Hodgson, M., R. Rempel, and S. Kennedy. 1999. Measurement and prediction of typical speech and background-noise levels in university classrooms during lectures. *J Acoust Soc Am* 105(1):226–233. DOI:10.1121/1.424600.

ISO 3382-1:2009. 2009. Acoustics—Measurement of room acoustic parameters. Part 1: Performance spaces. Geneva, Switzerland: International Organization for Standardization.

ISO 9921:2003. 2003. Ergonomics—Assessment of speech communication. Geneva, Switzerland: International Organization for Standardization.

Kosała, K. 2008. Global index of the acoustic quality of sacral buildings at incomplete information. *Arch Acoust* 33(2):165–183.

Kosała, K. 2011. A single number index to asses selected acoustic parameters in churches with redundant information. *Arch Acoust* 36(3):545–560. DOI:10.2478/v10168-011-0039-3.

Kosała, K. 2012. Singular vectors in acoustic simulation tests of St. Paul the Apostle Church in Bochnia. *Arch Acoust* 37(1):23–30. DOI:10.2478/v10168-012-0004-9.

Koszarny, K. 1992. Ocena hałasu szkolnego przez nauczycieli oraz jego wpływu na stan zdrowia i samopoczucie. [Noise assessment by school teachers and its impact on health and wellbeing]. *Roczniki Państwowego Zakładu Higieny* 43(2):201–210.

Kotus, J., M. Szczodrak, A. Czyżewski, and B. Kostek. 2010. Long-term comparative evaluation of acoustic climate in selected schools before and after acoustic treatment. *Arch Acoust* 35(4):551–564. DOI:10.2478/v10168-010-0042-0.

Leśna, P., and E. Skrodzka. 2010. Subjective evaluation of classroom acoustics by teenagers vs. reverberation time. *Acta Phys Pol A* 118(1):115–117. DOI:10.12693/APhysPolA.118.115.

Mikulski, W. 2012. Właściwości akustyczne sal lekcyjnych w szkołach podstawowych – Szacowanie wskaźnika transmisji mowy na podstawie czasu pogłosu. [Acoustic properties of classrooms in primary schools—Estimating speech transmission index from the reverberation time]. *Proceedings of the 59th Open Seminar on Acoustics Boszkowo*, Poland, September 10–14, 2012.

Mikulski, W., and J. Radosz. 2010. Wpływ objętości i wyposażenia na właściwości akustyczne sal lekcyjnych. [Effect of volume and equipment on acoustic performance of classrooms]. *Proceedings of the Noise Control 2010*, Książ, Poland, June 6–9, 2010.

Mikulski, W., and J. Radosz. 2011. Acoustics of classrooms in primary schools – Results of the reverberation time and the speech transmission index assessments in selected buildings. *Arch Acoust* 36(4):777–794. DOI:10.2478/v10168-011-0052-6.

ÖNORM B 8115-3:2005. 2005. Schallschutz und Raumakustik im Hochbau—Teil 3: Raumakustik [Austrian standard: Sound insulation and architectural acoustics in building construction—Part 3: Architectural acoustics]. Vienna, Austria: Austrian Standards.

Piechowicz, J. 2004. Global index of the acoustic climate. *Arch Acoust* 29(3):411–425.

Pleban, D. 1999. Computer simulation of the indices of the acoustic assessment of machines. *Arch Acoust* 24(4):443–453.

Pleban, D. 2010. Method of acoustic assessment of machinery based on global acoustic quality index. *Arch Acoust* 35(2):223–235. DOI:10.2478/v10168-010-0021-5.

Pleban, D. 2011. A global index of acoustic assessment of machines – Results of experimental and simulation tests. *Int J Occup Saf Ergon* 17(3):277–286. DOI:10.1080/10803548.2011.11076894.

PN-B-02151-02:1987. 1987. Akustyka budowlana. Ochrona przed hałasem pomieszczeń w budynkach. Dopuszczalne wartości poziomu dźwięku w pomieszczeniach [Polish standard: Building acoustics. Noise protection of apartments in buildings. Permissible values of sound pressure level]. Warsaw, Poland: Polski Komitet Normalizacyjny [Polish Committee for Standardization].

Radosz, J. 2012. Wpływ właściwości akustycznych sal lekcyjnych na poziom ciśnienia akustycznego mowy nauczycieli. [Effect of classroom acoustics on the sound pressure level of teachers' speech]. *Med Pr* 63(4):409–417.

Radosz, J. 2013. Global index of the acoustic quality of classrooms. *Arch Acoust* 38(2):159–168. DOI:10.2478/aoa-2013-0018.

Radosz, J., and W. Mikulski. 2012. Ocena właściwości akustycznych pomieszczeń pracy nauczycieli na przykładzie wybranych szkół podstawowych. [Evaluation of teachers' workplace acoustics based on selected primary schools]. *Bezpieczeństwo Pracy – Nauka i Praktyka* 6:16–19.

Rudno-Rudzińska, B., and K. Czajkowska. 2010. Analysis of acoustic environment on premises of nursery schools in Wrocław. *Arch Acoust* 35(2):245–252. DOI:10.2478/ v10168-010-0023-3.

Sato, H., and J. S. Bradley. 2008. Evaluation of acoustical conditions for speech communication in working elementary school classrooms. *J Acoust Soc Am* 123(4):2064–2077. DOI:10.1121/1.2839283.

SFS 5907:en. 2006. Acoustics classification of spaces in buildings. Helsinki, Finland: Finnish Standard Association.

7 Studies on Acoustic Properties of Open-Plan Office Rooms

Witold Mikulski

CONTENTS

7.1   INTRODUCTION

In office premises of open-plan space type known also as open-plan office rooms, noise is the most arduous factor in the work environment. Nonetheless, up until the year 2012, there was no harmonized and commonly accepted approach to assessment of acoustic properties of rooms of that kind. In 2012, parameters characterizing acoustic properties of rooms of that specific type have been defined in the standard EN ISO 3382-3:2012 which are: the distraction distance, the privacy distance, the A-weighted sound pressure level of speech at the distance of 4 m, and the speech spatial decay rate. The standard also proposes specific methods for determining values of the newly-defined quantities and the values considered acceptable for the so-called "good acoustic conditions." In the following, the requirements established in that standard will be referred to as "the additional requirements."

The requirements applicable in Poland are defined in Polish standard PN-B-02151-4:2015. The criteria should be considered the minimum requirements and therefore they will be further referred to as "the basic requirements" (cf. e.g., [Mikulski

2019b]). These basic requirements concerning reverberation properties of open-plan office rooms and the assessment parameters applicable to such spaces is equivalent to the "sound absorption area of the room" (in brief, the "room absorption") determined in octave frequency bands with mid-band frequencies of 0.5 kHz, 1 kHz, and 2 kHz (hereinafter referred to in short as the 0.5–2 kHz frequency band). On the other hand, Polish standard PN-B-02151-4:2015 specifies the minimum value for the room absorption of an open-plan office room as equaling 1.1 m^2 (standardized to the unit surface area of orthographic projection of the room) and a simple "engineering" method recommended to be used to calculate value of the quantity (by summing up sound absorption values of all structural components and equipment items of the room, a procedure called also the statistical method). The national standard can be therefore considered a minimum requirement and called "the basic requirement" as opposed to requirements established for "good acoustic conditions" defined in EN ISO 3382-3:2012 and called "the additional requirements."

The research work carried out worldwide on acoustic properties of open-plan office rooms and the acoustic environment prevailing in rooms of that type can be categorized into several currents. The first current encompasses studies concerning noise as a factor contributing to tiresomeness of work in the office environment [Evans and Johnson 2000; Kim and de Dear 2013; Haapakangas et al. 2014; Yadav et al. 2017]. The second trend is represented by publications dealing with analysis of phenomena and technical solutions affecting acoustic environment in open-plan offices [Yu et al. 2016; Mikulski 2016, 2019a; Vellenga-Persoon et al. 2017; Muller-Trapet and Gover 2019], including research carried out in laboratory conditions [Kostallari et al. 2017]. The third current in the literature of the subject is represented by empirical studies on acoustic properties of office rooms [Virjonen et al. 2009; Mikulski 2016, 2018, 2019a; Davidsson and Hodsman 2017; Haapakangas and Hongisto 2017; Vellenga-Persoon et al. 2017]. The fourth trend encompasses studies on open-plan office rooms carried out with the use of analytical methods based mainly on mathematical models implemented in the form of computer programs [Rindel 2012; Rindel and Christensen 2012; Keränen and Hongisto 2013; Smith 2017; Mikulski 2019a]. The fifth current is represented by studies on design work on acoustic treatment of the rooms in question [Kulowski 2011; Mikulski 2016, 2019b; Asselineau and Gaulupeau 2017]. The sixth current in the relevant literature covers deliberations on assessment criteria applicable to open-plan office rooms [Bradley et al. 1999; Ebissou et al. 2015; Yu et al. 2016; Vellenga-Persoon et al. 2017; Yadav et al. 2017; Cabrera et al. 2018; Mikulski 2018, 2019a, 2019b; Nurzyński 2018]. The seventh subject of the related studies consists in examining the effect of speech sound masking on the acoustic environment in such rooms [Edgington and Stevens 2017; Zaglauer et al. 2017; Renz et al. 2018]. Computational methods used in studies based on empirically verified simulations of the acoustic field in rooms (e.g., the second or the fourth trend or the fifth trend combined with the third trend) should be considered the most reliable ones and recommended as suitable tools to be used in the process of adaptation of open-plan office rooms to the purpose they are expected to serve—providing a friendly and comfortable environment for office work.

This chapter presents the process of adaptation of an open-plan office to the criteria set out in basic requirements (PN-B-02151-4:2015) and additional requirements

(EN ISO 3382-3:2012). The process takes into account the design methods based on simple calculation methods specified in PN-B-02151-4:2015 and computational methods of sound field simulation in rooms (combined ray-tracing and image-source geometrical methods implemented in ODEON software). The process included empirical verification of results obtained by numerical simulations before and after modernization (acoustic treatment). The chapter ends with a presentation of results of empirical investigations carried out in 15 open-plan type office rooms in Poland.

To ensure that acoustic properties in open-plan office rooms are on an acceptable level (basic requirements defined in PN-B-02151-4:2015 and, especially, additional requirements set out in EN ISO 3382-3:2012), it is necessary to employ a number of technical measures not applied to date in rooms of that type including, among other things, very large quantities of sound-absorbing materials (including suspended sound-absorbing ceilings) characterized by high sound absorption values, tall acoustic screens with high sound absorption, and in some cases, even speech masking sounds. Deployment of such technical measures in the rooms, despite their indispensability from the point of view of the imperative to achieve acceptable conditions of the acoustic environment, raises doubts among designers, administrators, owners, and users of such facilities. Apart from the necessity to finance additional investment into the room, the doubts concern firstly the risk of disruption to the current functionality of the rooms and secondly, actual effectiveness of the proposed solutions in achieving the required acoustic properties. This is a vital issue which hinders the process of improvement of work conditions in such rooms. In this chapter, the second of the issues is discussed in detail. The present author's own empirical studies and simulations carried out in the Central Institute for Labour Protection–National Research Institute in Warsaw, Poland, resulted in development of a toolbox of technical measures recommended to be used in projects aimed at adaptation of acoustic properties of open-plan office rooms to criteria set out in terms of both basic (PN-B-02151-4:2015) and additional (EN ISO 3382-3:2012) requirements.

7.2   ACOUSTIC CHARACTERIZATION OF OPEN-PLAN OFFICE ROOMS

The so-called open-plan office rooms are called also large-space rooms, large-area office rooms, or open-space rooms intended for office, intellectual, or administrative work. The rooms typically have large cubic capacity (cubature in the range 200–1000 m^3) and large surface area (typically in the range 70–300 m^2). In most cases, the height of a room of that type is relatively small (3–3.5 m). From about 10 to about 40–70 workstations are typically accommodated in such spaces (rooms for helpdesk consultants or telephone operators where more that 200 people work at the same time, constitute a separate group of open-plan workrooms to which other acoustical requirements apply and which will not be discussed in the following).

Typically, acoustic treatment of open-plan office rooms comprises installation of suspended sound-absorbing ceilings (from the point of view of acoustic properties, the suspension height is too small in the majority of cases due to small height of the rooms), and covering floors with fitted carpet. Only in very few cases, there are sound-absorbing materials placed on walls of the rooms. Some of the open-plan office rooms are equipped with acoustic screens. The screens are installed between individual workstations and sporadically, free-standing acoustic screens separate further these workstations from traffic routes demarcated in such premises. Screens are also mounted or placed on tops of work desks. The height of acoustic screens, in the majority of cases, does not exceed the height of a seated person. In very rare cases, acoustic screens are made of materials with sound-absorbing properties.

The specificity of work carried out in rooms of this type consists, among other things, in sporadic verbal communication between workers, printing texts, and sporadic making and/or answering telephone calls. The noise due to conversations (although also occurring sporadically) contributes to arduousness of the work environment to the largest degree. Apart from loudness, tiresomeness of these conversations is connected with the fact that being information-carrying messages, they may have a distracting effect on co-workers.

7.3 ACOUSTIC REQUIREMENTS APPLICABLE TO OPEN-PLAN OFFICE ROOMS

7.3.1 ACOUSTIC REQUIREMENTS CONCERNING THE ACOUSTIC ENVIRONMENT IN OPEN-PLAN OFFICE ROOMS

Acoustic requirements concerning the acoustic environment in open-plan office rooms refer to resultant noise level at workstations, acoustic properties of the rooms as the whole, and the background noise (from installed technical equipment and noise penetrating from outside into the room). The requirements established for open-plan office rooms, apart from limits set out for noise and reverberation, include provisions concerning necessary acoustic separation between individual workstations.

The parameters characterizing the acoustic environment in open-plan office rooms can be categorized into four groups:

A. Parameters characterizing noise at workrooms and in workplaces as per PN-B-02151-2:2018;
B. Parameters characterizing reverberation properties of the room (PN-B-02151-4:2015):
    1. the room absorption A (or, alternatively, the room absorption standardized to 1 m^2 of orthographic projection (surface area) of orthographic projection of the room $A_{1\,m^2}$), in short—"the room absorption standardized to 1 m^2 of the floor surface area," and
    2. the reverberation time T;
C. Parameters characterizing a room from the speech sounds propagation point of view (determined with the use of the reference speech sound source and the assumed A-weighted sound pressure level of background noise in the room; EN ISO 3382-3:2012):
    • distribution of speech transmission index values (in most cases, in a horizontal plane $STI(x, y)$) and, in particular, of STI values as functions of the distance from the reference speech sound source $STI(r)$, plus two parameters characterizing the quantity (at a given A-weighted sound pressure level) of background noise:

3. the distraction distance r_D and
4. the privacy distance r_P.
    –   distribution pattern of the A-weighted sound pressure level from the
        reference speech sound source in the room (in most cases, in a hori-
        zontal plane $L_{p,A,S}(x, y)$) and, in particular, functional dependence
        of the A-weighted speech sound pressure level on the distance from
        the reference speech sound source $L_{p,A,S}(r)$, plus two parameters
        characterizing that quantity:
5. the A-weighted sound pressure level of speech at the distance of 4 m
    (from the reference speech sound source) $L_{p,A,S,4\,m}$ and
6. the speech spatial decay rate $D_{2,S}$ (the difference in the A-weighted
    speech sound pressure level measured at distances of 1 and 2 m from
    the source).
D. Parameters characterizing the background noise include the A-weighted
    sound pressure level of background noise $L_{p,A,bgn}$ (including possible speech
    sound masking signals).

Although no criteria for the privacy distance r_P were established in additional
requirements defined in EN ISO 3382-3:2012, it seems to be a good idea to take the
parameter into account as its value can be used to compare different rooms from the
point of view of their acoustic properties.

For the purpose of the present study, acoustic properties of open-plan office
rooms will be characterized with the parameters listed above as items numbered 1
through 6.

7.3.2 BASIC REQUIREMENTS DETERMINED IN TERMS OF REVERBERATION TIME AND ROOM ABSORPTION

As the room absorption A depends (in an obvious way) on the room cubic capacity,
to make comparisons between rooms with different cubature more explicit as far
as their acoustic properties are concerned, the concepts of the room absorption A
standardized to 1 m² of orthographic projection of the room will be used in the
present chapter interchangeably. The latter is defined by the formula:

$$A_{1\,m^2} = \left\{ A / S_{floor} \right\} \, m^2 \qquad (7.1)$$

where:
   A (m²): room absorption
   S_{floor} (m²): room orthographic projection (floor, in most of cases) surface area

Table 7.1 summarizes the minimum acceptable value of the room absorption standardized to 1 m² of the orthographic projection of the room (i.e., to the unit floor area)
$A_{1\,m^2}$ applicable to open-plan office rooms according to Polish standard PN-B-02151-
4:2015 (the basic requirement).

TABLE 7.1

The Minimum Acceptable Value of the Room Absorption Standardized to 1 m² of Orthographic Projection of the Room

Room Type	$A_{1\,m^2,min\_accept}$ (m²)
Large-space offices, open-space rooms for administrative work called "open-plan offices", operation halls in banks and office, customer service bureaus, and other rooms with similar function	1.1

Note: $A_{1\,m^2,min\_accept}$ (in octave frequency bands with mid-band frequencies of 0.5 kHz, 1 kHz, and 2 kHz) calculated as per Polish standard PN-B-02151-4:2015 (the basic requirement).

With the minimum acceptable room absorption value given in Table 7.1 taken into account, the maximum acceptable reverberation time for the rooms in question can be calculated from the following approximate formula:

$$T_{max\_accept} \approx 0.161\{V\}/\{A_{min\_accept}\} = 0.161\{V\}/\{A_{1m^2,\ min\_accept} \cdot S_{floor}\}$$
$$= 0.161\{V\}/1.1\{S_{floor}\} \approx 0.146\{H\}\ \text{s} \tag{7.2}$$

where:

V (m³): the room volume (cubature)

A_{min_accept} (m²): the minimum acceptable room absorption

$A_{1\,m^2,min\_accept}$ (m²): the minimum acceptable room absorption standardized to 1 m² of orthographic projection of the room (floor surface area, see Table 7.1)

S_{floor} (m²): the room floor surface area

H (m): the room height

For instance, for rooms 3–3.5 m high, the maximum acceptable reverberation time value shall be in the range of 0.44–0.51 s.

7.3.3 ROOM ACOUSTICS REQUIREMENTS DETERMINED IN TERMS OF THE SPEECH TRANSMISSION INDEX

An important factor affecting concentration of the mind among workers is keeping trace (consciously or not) of conversations carried out by other individuals present in the same room. Therefore, in an open-plan office room it is necessary to see to it that the speech intelligibility is as low as possible. The related acoustical quantity is called the speech transmission index and denoted *STI*. Table 7.2 lists the relationship between values of the speech transmission index *STI* and the speech intelligibility determined subjectively according to EN 60268-16:2011 and EN ISO 9921:2003.

TABLE 7.2

The Relationship between Values of the Speech Transmission Index *STI* and the Speech Intelligibility Determined Subjectively

Speech Intelligibility	Bad	Poor	Average	Good	Excellent
Speech transmission index *STI*	0–0.3	0.3–0.45	0.45–0.6	0.6–0.75	0.75–1

Source: EN 60268-16:2011.

The minimum difference in speech intelligibility recognized aurally by a human (the so-called Just Notification Difference of Speech Transmission Index, *JND STI*) corresponds to the difference on value of the speech transmission index *STI* equaling 0.03 [Bradley et al. 1999]. Although the speech intelligibility is a quantity that can be determined between every two workstations, from the point of view of shaping acoustic conditions at workstations in a room, the more justified approach seems to be that based on parameters which characterize the averaged acoustic properties of the space in question. The parameters determining average acoustic properties of a room from the point of view of the speech intelligibility range are the distraction distance r_D and the privacy distance r_P. The two parameters determine at what distance (in meters) from the reference speech sound source (Table 7.3), the speech transmission index value equals 0.5 (r_D) and 0.2 (r_P), respectively. The lower are values of these parameters, the smaller will be the speech intelligibility zone in the room, which means that the area around a speaker in which his/her speech is intelligible will be smaller.

The "good acoustic conditions" (additional requirement as per EN ISO 3382-3:2012) label can be attributed to an open-plan office room when the value of the distraction distance r_D does not exceed 5 m (i.e., $r_{D,max\_accept} = 5$ m).

TABLE 7.3

The Emission Sound Pressure Level (the Sound Pressure Level in Open Space) at the Distance of 1 m from the Reference Omnidirectional Speech Sound Source

	Emission Sound Pressure Level (dB)							
	Octave Band Mid-Frequency (kHz)							
Source Type	0.125	0.25	0.5	1	2	4	8	A-weighted
Omnidirectional	49.9	54.3	58.0	52.0	44.8	38.8	33.5	57.4

Source: EN ISO 11201:2010, EN ISO 3382-3:2012.

Neither the standard EN ISO 3382-3 nor any other normative document issued until now (2019) establishes a maximum admissible value for the privacy distance r_P. For this reason, the parameter is used in the present study only for the purpose of making comparisons between rooms as far as their acoustic properties are concerned. Typical values of the privacy distance r_P in rooms of the open-plan type are greater than 20 m. As typical values of the distraction distance r_D according to EN ISO 3382-3:2012 are greater than 9–10 m, in the present chapter it is assumed that the value of the maximum acceptable privacy distance r_{P,max_accept} is twice the value of the maximum acceptable distraction distance r_{D,max_accept}:

$$r_{P,max\_accept} = 2 \cdot r_{D,max\_accept} = 2 \cdot 5 \text{ m} = 10 \text{ m} \tag{7.3}$$

where:

r_{D,max_accept} (m): the maximum acceptable distraction distance (5 m according to EN ISO 3382-3:2012)

r_{P,max_accept} (m): the maximum acceptable privacy distance

When determining the distraction distance r_D and the privacy distance r_P, the background noise is taken into account, whereas according to the definition proposed in the standard EN ISO 3382-3:2012, the background noise does not include the sounds of conversations held by people. Such approach may be considered controversial [Nurzyński 2018] as consideration of the effect of actual background noise (due to conversations) results, in the majority of cases, in improvement of acoustic conditions in the room, especially when assessment of the conditions is based on values of the distraction distance r_D and the privacy distance r_P. In the opinion of the present author, passing over the background noise due to conversations in the A-weighted sound pressure level is appropriate for at least three reasons. Firstly, these are the conditions concerning the background noise conditions for which the standard EN ISO 3382-3:2012 determines acceptable assessment results, therefore in case of taking into account the noise due to the presence of people as a background noise component, it would be necessary to change these values (in particular, the distraction distance and the privacy distance). Secondly, the A-weighted sound pressure level of conversations depends on so many unpredictable factors that taking them all into account a priori in the design work would be very difficult. Thirdly, the nature of the background, at the same measured A-weighted sound pressure level of background noise, can be different. That can be either a hubbub of voices, or a result of one or only a few but clearly comprehensible conversations. In the first case, the background noise would have a positive, and in the second instance a negative effect on acoustic work conditions.

7.3.4 Room Acoustics Requirements Determined in Terms of the A-weighted Sound Pressure Level of Reference Speech Source

In open-plan office rooms, it is necessary to ensure that the workers' speech audibility is kept at a level as low as possible. The speech audibility is determined by means of the A-weighted sound pressure level of speech. The quantity can be evaluated for verbal communication between any two workstations, but from the point of view of shaping acoustic conditions in a room as a whole, it seems to be more appropriate to use parameters which represent average properties of the whole workspace.

The attribute "good acoustic conditions" (the additional requirement as per EN ISO 3382-3:2012) may be assigned to an open-plan office room if the value of the speech spatial decay rate $D_{2,S}$ equals or exceeds 7 dB (i.e., $D_{2,S} \geq 7$ dB) and the value of the A-weighted sound pressure level of speech at the distance of 4 m from the reference speech sound source (cf. data in Table 7.3) $L_{p,A,S,4\,m}$ is less then or equals 48 dB (i.e., $L_{p,A,S,4\,m} \leq 48$ dB).

Typical values of the two parameters in open-plan office rooms (according to EN ISO 3382-3:2012) are as follows: the speech spatial decay rate $D_{2,S}$ less than 5 dB and the A-weighted sound pressure level of speech at the distance of 4 m $L_{p,A,S,4\,m}$ higher than 50 dB.

7.4   THE METHOD OF DETERMINING PARAMETERS CHARACTERIZING ACOUSTIC PROPERTIES OF OPEN-PLAN OFFICE ROOMS

7.4.1   THE REVERBERATION TIME MEASURING AND THE ROOM ABSORPTION CALCULATION METHODS

The parameters most commonly used to characterize acoustic properties of rooms (and thus also acoustic conditions prevailing therein) are the reverberation time T and the room absorption A. In most cases, the first of the parameters is determined with the use of measuring methods, whereas the second quantity is typically evaluated by way of simple calculation methods (such as the routine defined in PN-B-02151-4:2015). An approximate relationship between the reverberation time and the room absorption was quoted above as Equation 7.2.

7.4.1.1   The Reverberation Time

The reverberation time is determined typically in octave frequency bands with the mid-band frequencies of 0.125 kHz, 0.25 kHz, 0.5 kHz, 1 kHz, 2 kHz, 4 kHz, and 8 kHz and denoted T. In some applications, for the purpose of single-number characterization of the reverberation time, its value for 1 kHz octave band is used (T_{1kHz}) or the average of values for octave frequency bands centered at 0.5 kHz, 1 kHz, and 2 kHz (T_{mf}). In open-plan office rooms, by way of analogy to the room absorption, the most important are results of measurements taken for octave frequency bands with mid-band frequencies of 0.5 kHz, 1 kHz, and 2 kHz. The parameter is determined with the use of measurement methods defined in the standard EN ISO 3382-2:2008. The room absorption values are determined by averaging results of measurements carried out at microphone locations shown in Figure 7.1.

Open-plan offices are typically equipped with many irregularly distributed equipment items and at the same time, in many cases the whole space is divided

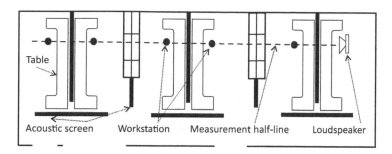

FIGURE 7.1 The measurement half-line on orthographic projection of the room with marked measurement points to determine the distraction distance and the privacy distance as well as the speech spatial decay rate and the A-weighted sound pressure level of speech at the distance of 4 m.

acoustically into smaller areas (for instance, by deploying acoustic screens), therefore the reverberation time values measured in different portions of the room may differ. In such case, besides the reverberation time characterizing the room globally, it may prove necessary to determine the reverberation time values for different portions of the room and, in extreme cases, the spatial distribution of the reverberation time.

7.4.1.2 The Room Absorption

The standard PN-B-02151-4:2015 establishes a computational method (based on the so-called statistical methods) for determining the room absorption value which depends on sound absorption properties of all surfaces (boundaries of the room), the sound absorption of equipment installed in the room, and the sound absorption in air. The room absorption is calculated with the use of the following formula:

$$A_{\text{room}} = A_{\text{surf}} + A_{\text{equip}} + A_{\text{air}} = \sum_{i=1}^{n} \alpha_i \cdot S_i + \sum_{j=1}^{q} A_{\text{equip},j} + 4mV \ \text{m}^2 \qquad (7.4)$$

where:

A_{surf} (m^2): the sound absorption of room surfaces (walls, floor, ceiling, etc.)
A_{equip} (m^2): the sound absorption of equipment items
A_{air} (m^2): the sound absorption resulting from sound damping in air
n : the number of surfaces in the room
α_i : the sound absorption coefficient for i-th surface in the room
S_i (m^2): the area of i-th surface in the room
q: the number of equipment items for which sound absorption was determined
$A_{\text{equip},j}$ (m^2): the sound absorption of j-th equipment item
m (Np/m): the sound power absorption coefficient in air in (cf. Table 7.4)
V (m^3): the room cubature

TABLE 7.4

Values of the Sound Power Absorption Coefficient in Air m (as per PN-B-02151-4:2015)

Air Temperature/ Relative Humidity	In Octave Frequency Bands with Mid-Band Frequency of (kHz)		
	0.5	1	2
	Sound Power Absorption Coefficient in Air (Np/m)		
20°C/30%–50%	0.0006	0.0010	0.0019
20°C/50%–70%	0.0006	0.0010	0.0017

According to the standard PN-B-02151-4:2015, the room absorption A is determined independently in three octave frequency bands with mid-band frequencies of 0.5 kHz, 1 kHz, and 2 kHz.

The above-described simple way of determining the room absorption is a method offering an approximate ("engineering") accuracy which can be successfully used in the practice of, first of all, the room acoustics design [Mikulski 2019b].

7.4.2 Methods for Determining the Speech Transmission Index, the Distraction Distance, and the Privacy Distance

One parameter used to determine speech intelligibility is the speech transmission index *STI*. The quantity is also used indirectly to characterize average acoustic properties of rooms (for the reference speech sound source and at a given background noise level).

The speech transmission index *STI* is an objective measure based on weighted contribution of a certain number of frequency bands contained in the frequency range specific for speech signals. The contributions are determined at an effective signal-to-noise ratio. With the properly selected sound waveform of the test signal, the effective signal-to-noise ratio can be taken into account collectively as distortions in the time and nonlinearity domain, as well as the background noise. Distortion in the time domain (reverberation, echoes, and automatic gain control) can reduce fluctuation of the speech signal and worsen intelligibility. According to the concept of the speech transmission index *STI*, the signal-to-noise ratio values in the range from −15 to +15 dB are linearly dependent on the speech intelligibility index in the range from 0 to 1. The speech transmission index measuring method is determined in the standard EN 60268-16:2011.

In open-plan office rooms it is necessary to assure that conversations held at or between individual workstations do not distract uninterested workers. There is therefore the need to maintain low speech intelligibility between individual workstations. Average acoustic properties of an open-plan office room are determined from functional relationships between the speech transmission index *STI* and the distance from the reference sound source. Based on the form of the function, two parameters characterizing acoustic properties of the room are determined: the distraction distance r_D and the privacy distance r_P. In the beginning, values of the speech transmission index STI are determined at workstations situated on a half line (ray) with its origin at a selected workstation (Figure 7.1) at which the reference speech sound source (a loudspeaker) is placed.

As values of the speech transmission index STI are determined at a finite number of points at workstations (the number recommended in EN ISO 3382-3:2012 is 6–10; cf. measurement line in Figure 7.1), both of the two parameters (the distraction distance and the privacy distance) are determined based on linear interpolation of measured speech transmission index values STI versus distance from the reference speech sound source (i.e., $STI = F_1(r) = a \cdot r + b$, where a and b are constants determined by linear interpolation of the speech transmission index STI measurement results, cf. Figure 7.2). The distraction distance r_D is the distance for which the value of function F_1 equals 0.5 (i.e., $0.5 = F_1(r_D)$), and the privacy distance r_P is the distance for which the value of the function equals 0.2 (i.e., $0.2 = F_1(r_P)$).

The method is set out in EN ISO 3382-3:2012.

7.4.3 Methods for Determining the A-weighted Sound Pressure Level of Speech at the Distance of 4 m and the Speech Spatial Decay Rate

From values of the A-weighted sound pressure level generated by the reference speech sound source on measurement lines (Figure 7.1), a function F_2 is determined representing the relationship between the A-weighted sound pressure and the distance to source with the use of logarithmic interpolation, $L_{p,A,S} = F_2(r) = a \cdot \log_{10}(r) + b$, where a and b are constants determined by interpolation of the A-weighted sound pressure level of speech measurement results, as shown in Figure 7.3.

Given the form of function F_2, two parameters characterizing average acoustic properties of an open-plan office room are determined. The first of the

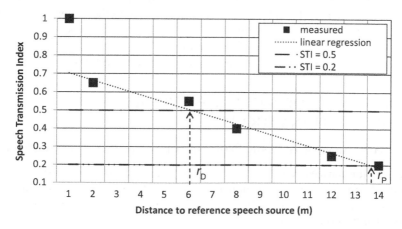

FIGURE 7.2 Linear interpolation of the speech transmission index STI values versus distance from the source, $STI = F_1(r)$, along the measurement half-line (Figure 7.1), the distraction distance r_D, and the privacy distance r_P.

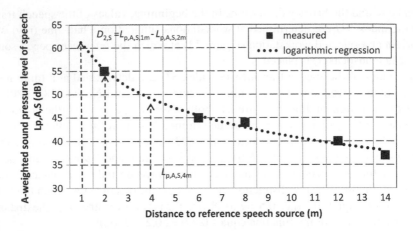

FIGURE 7.3 Logarithmic interpolation of the A-weighted sound pressure level of speech as a Function of distance from the reference speech sound source on the measuring half-line (Figure 7.1) $L_{p,A,S} = F_2(r)$, the speech spatial decay rate $D_{2,S}$, and the A-weighted sound pressure level of speech at the distance of 4 m $L_{p,A,S,4\,m}$.

parameters, namely the speech spatial decay rate $D_{2,S}$, is determined from the formula

$$D_{2,S} = L_{p,A,S,1m} - L_{p,A,S,2m} \;\; \text{dB} \tag{7.5}$$

where $L_{p,A,S,1\,m\,(2m)}$ (dB) is A-weighted sound pressure levels of speech read from the logarithmic interpolation line for the A-weighted sound pressure level for the distance of 1 m (2 m) from the reference speech sound source.

Value of the second parameter, namely the A-weighted sound pressure level of speech at the distance of 4 m $L_{p,A,S,4\,m}$, can be determined as the ordinate of the logarithmic interpolation line for the distance of 4 m from the reference speech sound source (i.e., $L_{p,A,S,4\,m} = F_2(4\,m)$) as shown in Figure 7.3.

The method of determining the parameters is defined in EN ISO 3382-3:2012.

7.5 ASSESSMENT OF ACOUSTIC PROPERTIES OF OPEN-PLAN OFFICE ROOMS

7.5.1 Assessment of Acoustic Properties of Fifteen Open-Plan Offices

Table 7.5 summarizes results of empirical studies concerning acoustical assessment of 15 office premises of open-plan type carried out by the present author and a research team of the Central Institute for Labour Protection–National Research Institute [Mikulski et al. 2016, 2019].

From the present author's own empirical studies on acoustic properties of 15 open-plan office rooms, it follows that the room absorption in about 26.7% of the examined rooms meets the basic requirements (in obligatory octave frequency bands 0.5 kHz,

TABLE 7.5

Results of Measurements and Assessment of Open-Plan Office Rooms According to Criteria Defined in Section 7.3

Criteria Parameters	Requirement	Room Number														
		1	2	3	4	5	6	7	8	9	10	11	12	13	14	15
$A_{1 m^2, 0.5\,kHz}$ (m²)	≥1.1	1.22	0.75	1.44	1.21	0.83	1.76	1.46	1.46	0.85	0.57	0.86	1.29	0.85	1.18	1.33
$A_{1 m^2, 1\,kHz}$ (m²)		1.11	0.63	1.19	1.18	0.79	1.69	1.58	1.39	1.04	0.71	0.82	1.11	1.08	1.19	0.99
$A_{1 m^2, 2\,kHz}$ (m²)		1.05	0.73	1.06	0.96	0.81	1.30	1.35	1.11	1.11	0.75	0.60	0.79	1.13	1.12	0.78
r_D (m)	≤5	28.4	22.2	12.3	7.6	12.5	6.7	10.9	12.0	9.8	4.9	7.3	9.3	10.2	8.2	11.3
$L_{p,A,S,4\,m}$ (dB)	≤48	52.2	51.9	49.5	50.0	53.1	45.6	45.9	45.2	53.9	55.4	53.4	50.9	51.0	46.8	50.6
$D_{2,S}$ (dB)	≥7	4.4	2.2	4.3	4.7	3.8	5.3	5.8	5.3	2.4	1.7	2.4	4.0	3.6	5.1	5.0

Source: Mikulski, W., Med. Pr., 67, 653–662, 2016; Mikulski, W. et al., Badania propagacji dźwięku i metod kształtowania warunków akustycznych w pomieszczeniach do pracy wymagającej koncentracji uwagi, Warszawa, CIOP-PIB 2016; Mikulski, W. et al., Badania oraz opracowanie metody kształtowania akustycznego środowiska pracy w wielkoprzestrzennych pomieszczeniach do pracy umysłowej, Warszawa, CIOP-PIB, 2019.

Notes:

$A_{1 m^2, 0.5\,(1, 2)\,kHz}$ (m²) room absorption standardized to 1 m² of its orthographic projection (floor surface area), in octave bands with mid-band frequencies of 0.5 kHz (1 kHz, 2 kHz) (the basic requirement)

r_D (m) distraction distance (additional requirement)

$L_{p,A,S,4m}$ (dB) A-weighted sound pressure level of speech at the distance of 4 m (additional requirement)

$D_{2,S}$ (dB) speech spatial decay rate (additional requirement)

light gray shading: the criterion is met

dark gray shading: the criterion is not met.

1 kHz, and 2 kHz as per PN-B-02151-4:2015), fulfills the basic requirements in the next 46.7% of the examined cases but only in some octave frequency bands, and in about 26.7% of the cases, the requirement is met in neither of the frequency bands. In the group of 7 rooms that meet the requirements for the room absorption in some frequency bands: in 5 of those rooms the condition is not met for the highest frequency (i.e., 2 kHz, which follows from low sound absorption values of sound-absorbing materials used in those rooms) and in 2 of those rooms, the condition is not met for the lowest frequency (0.5 kHz, which follows from the absence of the effect of a room absorption increase due to introduction of suspended ceiling).

In all of the analyzed open-plan office rooms, the additional requirements (defined for the so-called "good acoustic conditions" in EN ISO 3382-3:2012) are not met. In about 33% of the cases, additional requirements are met in respect to some parameters only. In particular, the distraction distance r_D met the criterion in 6.7% of cases (one room), and the A-weighted sound pressure level of speech at the distance of 4 m—in 26.7% of all cases (four rooms).

Based on the presented research results, it can be stated that additional requirements set out in the EN ISO 3382-3:2012 standard for the so-called "good acoustic conditions" are more demanding than the basic requirements established in PN-B-02151-4:2015.

7.5.2 Example Research Results and Acoustic Quality Assessments Concerning a Modern Open-Plan Office Room

The study was carried out in an up-to-date office facility in which numerous technical and organizational solutions were employed with the purpose of providing a good acoustic environment. Within the whole office facility area, open-plan office rooms, meeting rooms, rooms intended for carrying on telephone calls, rooms dedicated for work of confidential nature, printing and photocopying points, and welfare amenities have been sectioned off.

The examined open-plan office room has the form of a cuboid with dimensions 20.4 m × 4.5 m × 3.2 m, volume of 293.76 m³, and the room plan (i.e., the floor) surface area of 91.8 m². In the room there are sound-absorbing elements suspended under the ceiling, numerous sound absorbers, sound-absorbing sheet flooring, and a plurality of other solutions and equipment items affecting acoustic properties of the room (a sofa, lockers, soft seats, etc.). However, no acoustic screens were provided in the office space.

7.5.2.1 Basic Requirements

The reverberation time T of the room is shown in Figure 7.4. The maximum acceptable reverberation time in the 0.5–2 kHz band, calculated from Equation 7.2 (i.e., from the minimum acceptable room absorption set out in PN-N-02151-4:2015) was 0.47 s.

The reverberation time vs. frequency plot of the room is typical for spaces with suspended sound-absorbing ceilings (i.e., in the 0.5–2 kHz band, the reverberation time increases with increasing frequency). The feature, on one hand, evidences too

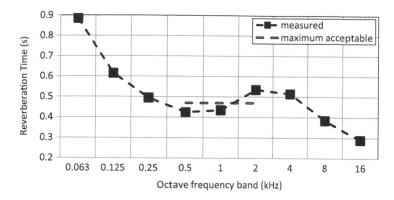

FIGURE 7.4   The reverberation time measured in the examined room.

small room absorption for the frequency of 2 kHz, but on the other hand, it shows that the satisfactory room absorption observed in the 0.25–1 kHz band can be attributed to the fact of installing suspended sound-absorbing elements.

Figure 7.5 is a plot of the room absorption standardized to 1 m² of its orthographic projection calculated from the measured reverberation time according to Equations 7.2 and 7.1 and compared to the minimum acceptable value of the parameter defined in PN-B-02151-4:2015.

Based on the comparison illustrated in the above figure it can be found that in the octave band with the mid-band frequency of 2 kHz, the room absorption of the room

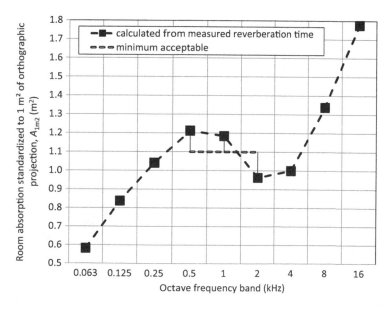

FIGURE 7.5   The room absorption standardized to 1 m² of orthographic projection of the room.

is insufficient (too small by about 13.4%, i.e., by 13.27 m²). A-weighted sound pressure levels of background noise in the room $L_{p,A,bgn}$ are diversified and assume values from 32.9 to 38.7 dB. Adopting the value of 40 dB established for the A-weighted sound pressure level of background noise in the standard PN-B-0251-2:2018 as the maximum acceptable one, it can be concluded that the background noise in the examined room does not exceed the acceptable level.

7.5.2.2 Additional Requirements

Measurements were carried out along a single measurement line for two sound propagation directions, of which the first direction corresponded to the reference speech sound source placed in the left portion of the room, and the second (opposite) direction was adopted for measurements taken when the reference speech sound source was situated near the right-hand-side end of the room (as shown in Figure 7.1).

Figure 7.6 shows results of measurements of the speech transmission index STI as a function of distance from the reference speech sound source on the measurement line in the "first" direction "↑". According to rules defined in EN ISO 3382-3, values of the speech transmission index STI at three points were rejected as they failed to meet the distance criterion (6 points were taken into account in calculations). The calculated value of the distraction distance r_D equals 8.2 m, and that of the privacy distance r_P equals 17.6 m. Value of the speech transmission index STI at the workstation closest to the reference speech sound source $STI_{nearest}$ equals 0.69.

Figure 7.7 illustrates results of measurements of the speech transmission index STI as a function of distance from the reference speech sound source in the opposite

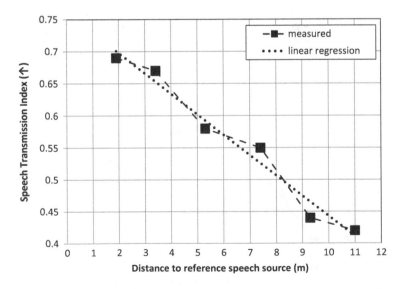

FIGURE 7.6 Result of measurements of the speech transmission index value STI versus distance from the reference speech sound source in the "first" direction "↑" (the fine dotted line represents linear regression of measurement results).

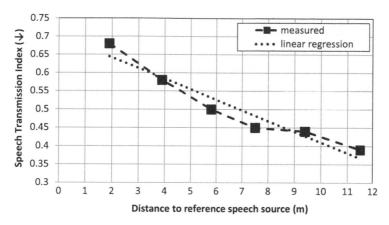

FIGURE 7.7 Result of measurements of the speech transmission index value *STI* versus distance from the reference speech sound source in the opposite direction "↓" (the fine dotted line represents linear regression of measurement results).

direction "↓". The calculated value of the distraction distance r_D equals 6.9 m, and that of the privacy distance r_P equals 17.2 m. Value of the speech transmission index at the workstation nearest to the reference speech sound source $STI_{nearest}$ equals 0.68.

Figure 7.8 shows a plot representing results of measurements of the A-weighted sound pressure level of speech versus distance from the reference speech sound source on the measurement line in the "first" direction "↑". The calculated A-weighted sound pressure level of speech value at the distance of 4 m $L_{p,A,S,4\,m}$ equals 50.7 dB, and value of the speech spatial decay rate $D_{2,S}$ equals 4.4 dB.

Figure 7.9 shows results of measurements of the A-weighted sound pressure level of speech versus distance from the reference speech sound source on the measurement line in the opposite direction "↓". The calculated A-weighted sound pressure

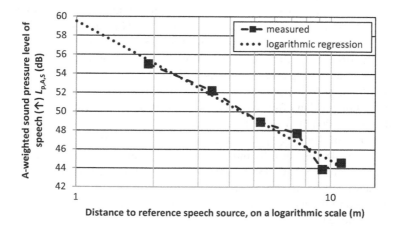

FIGURE 7.8 Results of measurement of A-weighted sound pressure level of speech versus distance from the reference speech sound source on the measurement line in the "first" direction "↑" (the fine dotted line represents linear regression of measurement results).

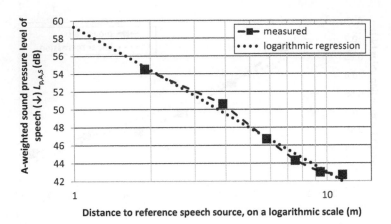

FIGURE 7.9 A-weighted sound pressure level of speech versus distance from the reference speech sound source on the measurement line in the opposite direction "↓" (the fine dotted line represents linear regression of measurement results).

level of speech value at the distance of 4 m $L_{p,A,S,4\,m}$ equals 49.5 dB, and value of the speech spatial decay rate $D_{2,S}$ equals 4.9 dB.

7.5.2.3 Summary of the Assessment

Values of parameters characterizing acoustic properties of an open-plan office room synthetically are given in Table 7.6.

The room does not meet the basic requirements determined in terms of the room absorption (as per PN-B-02151-4:2015). In particular, in the octave frequency band with the mid-band frequency of 2 kHz, the room absorption is 12.9 m² too small.

The room also fails to meet the additional requirements for the so-called "good acoustic conditions" according to criteria defined in EN ISO 3382-3. In particular, none of the parameters established in the standard achieves the required value. Value of the distraction distance r_D (the average of measurements taken for both of the two sound propagation directions, "↑" and "↓") is 7.55 m and fails to meet the criterion $r_D \leq 5$ m but exceeds the maximum acceptable value less than in other typical open-plan office rooms (where it is 9–10 m according to the standard EN ISO 3382-3:2012). Value of the A-weighted sound pressure level of speech at the distance of 4 m $L_{p,A,S,4\,m}$ (averaged) is 50.1 dB and does not meet the criterion $L_{p,A,S,4\,m} \leq 48$ dB but falls into the range of values typical for similar rooms (49–50 dB as per EN ISO 3382-3:2012). Value of the speech spatial decay rate $D_{2,S}$ (average) is 4.65 dB, does not meet the criterion $D_{2,S} \geq 7$ dB, and is even too low compared to the values characterizing other similar rooms (5–6 dB as per EN ISO 3382-3:2012).

To sum up, the examined open-plan office room meets neither the basic requirements set out in the standard PN-B-02151-4:2015 nor the additional requirements

TABLE 7.6

Results of Measurements and Acceptable Values of Parameters Used for the Purpose of Acoustic Assessment of Open-Plan Office Rooms Set Out in Basic and Additional Requirements

Parameter	Mid-Band Frequency (kHz)	Requirement	Measurement Results
Reverberation time (s)	0.5	≤ 0.47	0.43
	1	≤ 0.47	0.44
	2	≤ 0.47	0.54
Room absorption	0.5	≥ 1.1	1.2
standardized to 1 m² of the	1	≥ 1.1	1.2
floor surface area, $A_{1\,m^2}$ (m²)	2	≥ 1.1	1.0
A-weighted sound pressure level of background noise at workstations $L_{p,bgn,A}$ (dB)		≤ 40	32.9–38.7
A-weighted sound pressure level of speech at the distance of 4 m $L_{p,A,S,4\,m}$ (dB)		≤ 48	49.5–50.7
Speech spatial decay rate $D_{2,S}$ (dB)		≥ 7	4.4–4.9
Distraction distance r_D (m)		≤ 5	6.9–8.2
Privacy distance r_P (m)		$\leq 2 \cdot r_D$	17.2–17.6
Speech transmission index $STI_{nearest}$		—	0.68–0.69

Source: PN-B-02151-4:2015, EN ISO 3382-3:2012.
Note: Light gray shading: the criterion is met.
        Dark gray shading: the criterion is not met.

defined in EN ISO 3382-3:2012. However, organizational and technical measures implemented in the room do make it stand out among other examined open-plan offices as far as acoustic properties are concerned.

7.5.3 EXAMPLE RESULTS OF EXAMINATION AND ASSESSMENT OF OPEN-PLAN OFFICE ROOMS IN CONNECTION WITH THE EMPLOYED ACOUSTIC TREATMENT MEASURES

Table 7.7 summarizes the results of examination for 6 out of 15 open-plan office rooms listed in Section 7.5.1. The table includes also information about acoustic treatment measures deployed in the rooms: suspended sound-absorbing ceilings, sound-absorbing materials on walls, and acoustic screens [Mikulski et al. 2016, 2019].

As the measurements were taken in different rooms, comparative qualitative analysis was adopted as the research tool. Results of the analysis indicate that the open-plan office room with acoustic treatment of ceiling and walls meets the basic

TABLE 7.7

The Effect of Acoustic Treatment Measures on Acoustic Properties of Six Open-Plan Office Rooms with the Cubic Capacity from 295 m³ to 664 m³—Results of Measurements

Room No.	Sound-Absorbing Material Ceiling	Walls	Sound-Absorbing Acoustic Screens At Individual Workstations	Additional (For Groups of Workstations)	$A_{1\,m^2}$ (m²) 0.5 kHz	1 kHz	2 kHz	r_D (m)	$L_{p,A,S,4\,m}$ (dB)	$D_{2,S}$ (dB)
					Criterion					
					1.1	1.1	1.1	≤ 5	≤ 48	≥ 7
1	Applied	Applied	Applied	Applied	1.46	1.58	1.35	10.9	45.9	5.8
2	Applied	Applied	Applied	—	1.46	1.39	1.11	12.0	45.2	5.3
3	Applied	—	Applied	—	1.33	0.99	0.78	11.3	50.6	5.0
4	—	—	—	—	0.83	0.79	0.81	12.5	53.1	3.8
5	—	—	Applied	—	0.85	1.04	1.11	9.8	53.9	2.4
6	—	—	Applied	Applied	0.85	1.08	1.13	10.2	51.0	3.6

Source: Mikulski, W., Med. Pr., 67, 653–662, 2016; Mikulski, W. et al., Badania propagacji dźwięku i metod kształtowania warunków akustycznych w pomieszczeniach do pracy wymagającej koncentracji uwagi, Warszawa, CIOP-PIB 2016; Mikulski, W. et al., Badania oraz opracowanie metody kształtowania akustycznego środowiska pracy w wielkoprzestrzennych pomieszczeniach do pracy umysłowej, Warszawa, CIOP-PIB, 2019.

Notes:
$A_{1\,m^2}$ (m²) room absorption standardized to 1 m² of orthographic projection of the room
r_D (m) distraction distance
$D_{2,S}$ (dB) speech spatial decay rate
$L_{p,A,S,4\,m}$ (dB) A-weighted sound pressure level of speech at the distance of 4 m
— measure not applied (not installed)
light gray shading: the criterion is met
dark gray shading: the criterion is not met

criterion determined in terms of the room absorption (and thus also the reverberation time) of the room. By applying the suspended sound-absorbing ceiling alone, it is impossible to obtain adequate room absorption (the basic requirement). None of the rooms meet the additional requirements. Installation of the suspended sound-absorbing ceiling, as well as sound-absorbing materials on walls of the room, result in fulfillment of the requirement concerning the A-weighted sound pressure level of speech at the distance of 4 m, while achieving distinctly higher (yet still insufficiently high) value of the speech spatial decay rate $D_{2,S}$. Other relationships cannot be unambiguously identified and therefore it is necessary to undertake numerical simulation studies in order to determine the effect of minor differences in acoustic treatments on acoustic properties of rooms determined in terms of the above-mentioned acoustic parameters.

7.6 APPLICATION OF ODEON SIMULATION SOFTWARE IN DEVELOPMENT OF AN ACOUSTIC TREATMENT DESIGN FOR AN OPEN-PLAN OFFICE ROOM

7.6.1 INTRODUCTION—THE PURPOSE AND SCOPE OF THE STUDY

The objective of an acoustic treatment project (in the meaning applicable to this chapter) is to adjust acoustically an open-plan office room to meet specific requirements. To this end, it is necessary to undertake examination of acoustic properties of the room which will be further taken into account in the acoustic treatment project. To effect the examination, it is necessary to make a prediction (a prognosis concerning time-averaged values) of acoustic properties of the room which will enable to work out a room design meeting specifically imposed acoustic criteria. To make such a prediction for open-plan office rooms, it is necessary to employ advanced computational methods of determining sound field in rooms which are typically based on the geometrical sound field prediction approach (ray-tracing or image sources models). Due to the degree of complexity of the involved calculation algorithms, their numerical implementations take usually the form of dedicated computer programs for practical applications. In the present study of acoustic properties of the open-plan office room, the well-known ODEON software was used.

Four stages were planned within the framework of the study:

- **Stage I**—examination of acoustic properties of the room before acoustic treatment, comprising:

    Sub-stage Ia—measurements of acoustic properties of the existing room before realization of the acoustic treatment design and determination, with the use of a simple computational method (given in PN-B-02151-4:2015), how and by how much should the room absorption value be increased in order to meet the basic requirement determined in terms of the room absorption (the "basic" acoustic treatment project, i.e., meeting requirements concerning the room absorption value), and

    Sub-stage Ib—implementation of the room in a calculation computer program (including calibration of model and digital representation of the room) and numerical evaluation of acoustic properties of the room before acoustic treatment (numerical studies performed in ODEON software; a comparison of results of calculations and measurements);

- **Stage II**—examination of acoustic properties of the room after "basic" acoustic treatment (with suspended sound-absorbing elements), comprising:

    Sub-stage IIa—computational analysis of the solution proposed for the basic acoustic treatment, and

    Sub-stage IIb—measurements of acoustic properties of the room after realization of the proposed basic acoustic treatment, comparison between measurement and calculation results (verification of project results after realization of the basic acoustic treatment);

- **Stage III**—computational analysis of the effect of additional acoustic treatment elements (additional acoustic treatment) aimed at meeting the additional requirements (according to EN ISO 3382-3) for the room; and
- **Stage IV**—computational analysis of the effect of modification concerning some elements of acoustic treatment (basic and additional) on acoustic properties of the room.

7.6.2 Description of the Study Object

Empirical studies and numerical modeling were carried out for an existing open-plan office room. A view of the room is shown in Figure 7.10 (and in Figure 7.13). The room had the shape close to a prism with a base in the form of a trapezium. Dimensions of the room were 12.3–12.8 m × 5.85 m × 3 m, at the cubature of 220.3 m^3 and the floor surface area of 73.4 m^2. The room floor was covered with a fitted carpet with the weighted sound absorption coefficient $\alpha_w = 0.25$. Twelve workstations were arranged in that office space. Walls of the room were made of plastered ceramic bricks. Figure 7.11 depicts values of the sound absorption coefficient for the elements of which the room was constructed and sound absorption coefficient values for surfaces of individual room equipment items. Figure 7.12 shows a plot of background noise sound pressure levels in octave frequency bands (the plot was generated in the ODEON software, where the same background noise was used for the purpose of calculations).

7.6.3 Empirical Studies on Acoustic Properties of the Room Before Basic Acoustic Treatment—the Basic Acoustic Treatment Design

Sub-stage Ia included empirical examination of acoustic properties of the room without any acoustic treatment. The reverberation time of the room was measured and the room absorption was calculated (basic requirements) as well as the distraction

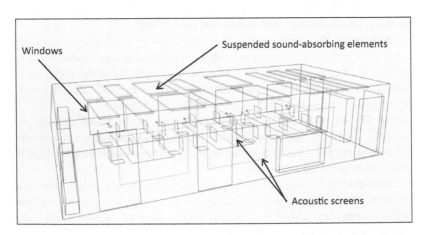

FIGURE 7.10 A schematic view of the room (generated in ODEON software).

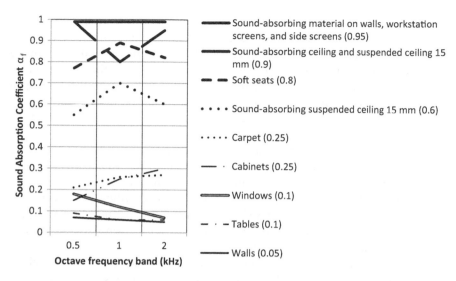

FIGURE 7.11 Sound absorption coefficients α of elements of which the examined room is constructed and sound absorption coefficient values of surfaces of various room equipment items (data from the ODEON software database and specifications made available by Ecophon company; in brackets, weighted values of the sound absorption coefficient are given).

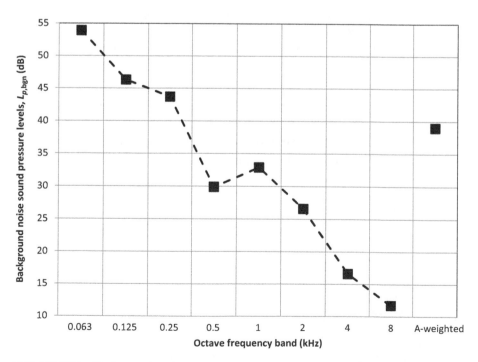

FIGURE 7.12 Background noise: sound pressure levels in octave frequency bands and A-weighted sound pressure level.

distance, the privacy distance, the A-weighted sound pressure level of speech at the distance of 4 m, and the speech spatial decay rate (additional requirements). A simple computational method (described in PN-B-02151-4:2015) was used to determine by how much the room absorption must be increased to meet the basic requirements. Moreover, a design of acoustic treatment called "basic" was also developed (the quantity, acoustical properties, and arrangement details of sound-absorbing elements suspended under the ceiling were determined). Results of the measurements are summarized in Table 7.8.

The obtained values of the room absorption A for different frequencies from the range 0.5–2 kHz are included in the range 46.4–54.5 m^2 (when standardized to 1 m^2 of orthographic projection of the room, they are included in the range 0.63–0.74 m^2) and are too small compared to the required values, as according to PN-B-02151-4:2015, they should be greater than 80.74 m^2 (standardized to 1 m^2 of orthographic projection of the room should be not less than 1.1 m^2). It is therefore necessary to increase the room absorption by about 26.1–34.3 m^2.

The (basic) acoustic treatment design provided for installation of sound-absorbing elements suspended under the ceiling with the weighted sound absorption coefficient 0.6 (the ceiling marked "0.6" in Figure 7.11) was taken into account. Twenty (20) elements with dimensions 0.6 m × 2.4 m × 0.04 m suspended 0.15 m under the ceiling structure were included in the acoustic design project (Figure 7.10). The increase of the room absorption value resulting from introduction of these elements (with only one side of each element taken acoustically into account) for frequencies 0.5 kHz, 1 kHz, and 2 kHz turned out to be 25.9 m^2, 27.9 m^2, and 27.9 m^2, respectively. It was estimated, in accordance with PN-B-02151-4:2015, that after the acoustic treatment, the room absorption should be about 80.5 m^2, 74.3 m^2, and 81.6 m^2 (or when standardized to 1 m^2

TABLE 7.8
Results of Measurements of Parameters Characterizing Acoustic Properties of the Room Before Acoustic Treatment

				In Octave Band with Mid-Band Frequency of (kHz)								
r_D (m)	r_P (m)	$L_{p,A,S,4 m}$ (dB)	$D_{2,S}$ (dB)	0.5	1	2	0.5	1	2	0.5	1	2
				T (s)			A (m^2)			$A_{1 m^2}$ (m^2)		
22.2	54.8	51.9	2.2	0.65	0.77	0.66	54.6	46.4	53.7	0.74	0.63	0.73

Notes:

r_D (m) distraction distance

r_P (m) privacy distance

$L_{p,A,S,4 m}$ (dB) A-weighted sound pressure level of speech at the distance of 4 m

$D_{2,S}$ (dB) speech spatial decay rate

T (s) reverberation time of the room

A (m^2) room absorption calculated from the measured reverberation time T of the room

$A_{1 m^2}$ (m^2) room absorption standardized to 1 m^2 of orthographic projection (floor surface area) of the room

of orthographic projection of the room, 1.10 m², 1.01 m², and 1.11 m², respectively). From this it follows that in octave frequency bands with mid-band frequencies 0.5 kHz and 2 kHz, the room absorption should be sufficient. For the frequency 1 kHz, according to calculations, it will be too small by about 7.9% (i.e., by 6.4 m²). One should therefore take additionally into account, for instance, the above-mentioned elements (4 pieces). However, in view of the fact that experimental feedback studies were planned to be performed after realization of the acoustic treatment work, such additional measure was not implemented in that stage of development of the basic acoustic treatment design. The rationale behind the decision included two arguments. Firstly, it can be reasonably expected that the actual sound absorption of the above-mentioned suspended sound-absorbing elements will be higher than that calculated above, as the elements will absorb the sound energy also with their peripheral sides (with the combined sound absorption of element sides of 5 m²) and, to some extent, even with surfaces of the elements facing the ceiling (at the suspension distance of 0.15 m). Secondly, one can expect that for the frequencies 0.5–1 kHz and at given element suspension distance, a resonance will occur resulting in an increase of sound absorption of the elements in that frequency range (by analogy to Figure 7.5). Therefore, actually, also for the frequency 1 kHz, the room absorption may prove to be adequate (the fact being verified by measurements presented in Sections 7.6.5 and 7.6.6).

As was already noted, the additional requirements set out in EN ISO 3382-3 standard are not taken into account in this stage of the design work and embodiment of the acoustic treatment project, as the indicators entail the scope of the basic acoustic treatment. It can be however noticed (from results of measurements shown in Table 7.8) that the distraction distance r_D of 22.2 m is too large (the maximum value acceptable for the so-called "good acoustic conditions" is 5 m), value of the speech spatial decay rate $D_{2,S}$ which equals 2.2 dB is very small (the minimum value acceptable for the so-called "good acoustic conditions" is 7 dB), and the A-weighted sound pressure level of speech at the distance of 4 m $L_{p,A,S,4\,m}$ equaling 51.9 dB is too high (the maximum value acceptable for the so-called "good acoustic conditions" is 48 dB). Value of the privacy distance r_P is 54.8 m.

In view of the above it must be concluded that acoustic properties of the room without acoustic treatment are unfavorable from the point of view of work in such an open-plan office room and fail to meet both the basic requirements set out in PN-B-02151-4:2015 and the additional requirements of EN ISO 3382-3:2012.

7.6.4 COMPUTATIONAL STUDIES PRECEDING REALIZATION OF BASIC ACOUSTIC TREATMENT—IMPLEMENTATION OF THE ROOM IN NUMERICAL SIMULATION PROGRAM AND SIMULATION RESULTS

The examined room (cf. data quoted in Section 7.6.2) was implemented in ODEON software (Figure 7.13) where based on data obtained from measurements, the model and its implementation were subject to verifcation and calibration.

Table 7.9 summarizes results of calculations (carried out with the use of ODEON software) and measurements (from Table 7.8) for parameters characterizing acoustic properties of the examined open-plan office room.

FIGURE 7.13   A view of room without acoustic treatment (generated in ODEON software).

TABLE 7.9
Results of Calculations (Carried Out with the Use of ODEON Simulation Software) and Results of Measurements of Parameters Characterizing Acoustic Properties of an Open-Plan Office Room Without Acoustic Treatment

Research Method	r_D (m)	r_P (m)	$L_{p,A,S,4m}$ (dB)	$D_{2,S}$ (dB)	0.125	0.25	0.5	1	2	4	8
					\multicolumn{7}{c}{Reverberation Time T (s)}						
Measurement	22.2	54.8	51.9	2.2	0.68	0.74	0.65	0.77	0.66	0.56	0.46
Calculation	21.04	55.45	55.31	2.21	0.68	0.74	0.64	0.62	0.63	0.57	0.44

In Octave Band with Mid-Band Frequency of (kHz)

Notes:
r_D (m) the distraction distance
r_P (m) the privacy distance
$L_{p,A,S,4m}$ (dB) the A-weighted sound pressure level of speech at the distance of 4 m
$D_{2,S}$ (dB) speech spatial decay rate
T (s) reverberation time in octave frequency bands

Figures 7.14 and 7.15 illustrate results of calculations concerning the speech transmission index *STI* at workstations and the A-weighted sound pressure level of speech at workstations.

Results of measurements and calculations concerning the reverberation time are illustrated in Figure 7.16.

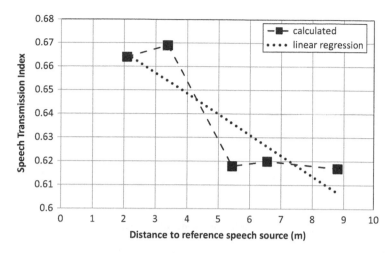

FIGURE 7.14   For the room without acoustic treatment, results of calculations (carried out with the use of ODEON software): the speech transmission index STI at workstations (the distraction distance r_D is 21.04 m and the privacy distance r_P is 55.45 m).

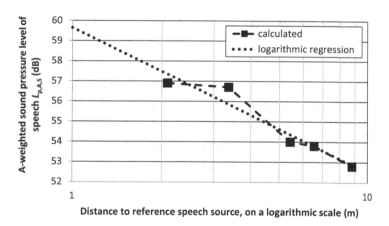

FIGURE 7.15   For the room without acoustic treatment, results of calculations (carried out with the use of ODEON software) for the A-weighted sound pressure level of speech at workstations (the A-weighted sound pressure level of speech at the distance of 4 m $L_{p,A,S,4\,m}$ is 55,31 dB and the speech spatial decay rate $D_{2,S}$ is 2,21 dB).

Relative differences between values of the parameters characterizing acoustic properties of the examined room obtained from measurements and simulations for the room without acoustic treatment are summarized in Table 7.10.

The relative error of the reverberation time for the room without acoustic treatment is included in the range from −4.5% to +1.8% and is acceptable (except for the frequency 1 kHz for which $\delta T = -19.5\%$). It is usually assumed that the significant

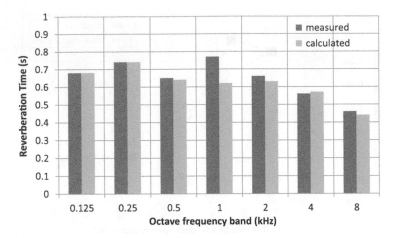

FIGURE 7.16 For the room without acoustic treatment, the reverberation time values obtained from measurements and numerical simulations.

TABLE 7.10

Relative Differences Between Values of Parameters Characterizing Acoustic Properties of the Examined Room Without Acoustic Treatment Obtained from Measurements and Calculations

				In Octave Band with Mid-Band Frequency of (kHz)						
				0.125	0.25	0.5	1	2	4	8
δr_D (%)	δr_P (%)	$\delta L_{p,A,S,4\,m}$ (%)	$\delta D_{2,S}$ (%)				δT (%)			
−5.2	1.2	6.6	0.5	0.0	0.0	−1.5	−19.5	−4.5	1.8	−4.3

Notes:

δr_D (%) relative difference between results of measurements and numerical simulations for the distraction distance

δr_P (%) relative difference between results of measurements and numerical simulations for the privacy distance

$\delta L_{p,A,S,4\,m}$ (%) relative difference between results of measurements and numerical simulations for the A-weighted sound pressure level of speech at the distance of 4 m

$\delta D_{2,S}$ (%) relative difference between results of measurements and numerical simulations for the speech spatial decay rate

δT (%) relative difference between results of measurements and numerical simulations for the reverberation time in individual octave bands

difference between reverberation time values obtained for 1 kHz (0.77 s from measurements and 0.62 s from calculations) is not an effect of properties of the sound-absorbing materials utilized (the course of the reverberation time vs. frequency plot is typical; cf. Figure 7.16), but rather a result of resonance phenomena occurring in the room (the effect not taken into account in these simulations). The phenomenon is clearly visible in the reverberation time chart obtained from measurements shown in Figure 7.16.

From the point of view of practice, the discrepancy between results of measurements and numerical simulations at that very frequency can be nevertheless accepted as one should expect that the sound absorption of suspended sound-absorbing elements will be higher (according to what was said in the preceding section) than that predicted based solely on properties of the type of sound-absorbing material utilized. The effect of increased room absorption due to suspending the sound-absorbing elements should compensate for the longer reverberation time in that frequency band (the thesis will be subject to verification by measurements in Section 7.6.6).

In view of the above one can accept the adopted computational method as a useful tool in the design studies on acoustic treatment of open-plan office rooms.

Figure 7.17 illustrates results of simulations for the spatial distribution of the speech transmission index *STI*, whereas Figure 7.18 shows simulation results for distribution of A-weighted sound pressure level of speech values in the room without acoustic treatment. It follows from those distribution patterns that at the height of a seated person's head ($h = 1.2$ m), the speech intelligibility is good in the whole room which is an undesirable feature in an open-plan office room (i.e., the value of the speech transmission index *STI* exceeds 0.6, and the A-weighted sound pressure level of speech value exceeds 53 dB, compared to the A-weighted sound pressure level of background noise of 39 dB).

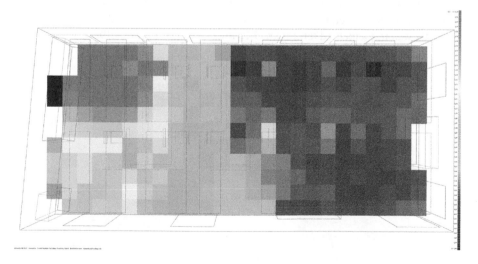

FIGURE 7.17 Distribution of speech transmission index *STI* values in horizontal plane ($h = 1.2$ m) of the room without acoustic treatment (numerical simulation results by ODEON software).

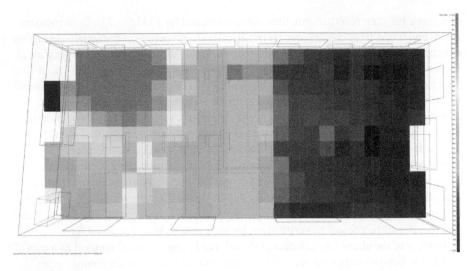

FIGURE 7.18 Distribution of A-weighted sound pressure level of speech values (denoted SPL(A)) in horizontal plane ($h = 1.2$ m) of the room without acoustic treatment (numerical simulation results by ODEON software).

7.6.5 Computational (Simulation) Studies on Acoustic Properties of a Room with the Proposed Basic Acoustic Treatment and Suspended Sound-Absorbing Elements Taken into Account

In the ODEON software, numerical simulation of the room acoustics with the proposed basic acoustic treatment which included the use of suspended sound-absorbing elements was implemented (cf. Figure 7.10 and a view of Figure 7.19).

FIGURE 7.19 A view of acoustic treatment of the room with suspended sound-absorbing elements taken into account (generated in ODEON software).

TABLE 7.11

Results of Numerical Simulations Aimed at Determining the Parameters Characterizing Acoustic Properties of the Room After Basic Acoustic Treatment with Suspended Sound-Absorbing Elements Taken into Account

Acoustic Treatment Measure	r_D (m)	r_P (m)	$L_{p,A,S,4\,m}$ (dB)	$D_{2,S}$ (dB)	In Octave Band with Mid-Band Frequency of (kHz)					
					0.5	1	2	0.5	1	2
					Reverberation Time (s)			Room Absorption (m²)		
Suspended sound-absorbing elements	13.12	26.76	51.9	3.62	0.49	0.5	0.51	72.4	70.9	69.5

Notes:
r_D (m) distraction distance
r_P (m) privacy distance
$L_{p,A,S,4\,m}$ (dB) A-weighted sound pressure level of speech at the distance of 4 m
$D_{2,S}$ (dB) speech spatial decay rate
T (s) reverberation time (for frequencies 0.5 kHz, 1 kHz, and 2 kHz)
A (m²) the room absorption (calculated from the reverberation time T)

Values of parameters characterizing acoustic properties of the examined room after basic acoustic treatment with suspended sound-absorbing elements taken into account obtained from calculations are presented in Table 7.11 as well as in Figures 7.20 and 7.21.

The values of the room absorption calculated for frequencies from the range of 0.5–2 kHz are included in the range 69.5–72.4 m² (or 0.95–0.99 m² when standardized to 1 m² of orthographic projection of the room) and are much larger that those without suspended sound-absorbing elements (without acoustic treatment) but still too small. According to PN-B-02151-4:2015, they should be larger than 80.7 m². It is therefore necessary to increase the room absorption value by some 10%–14%.

Figures 7.22 and 7.23 present a representation of simulation results for distribution of the speech transmission index *STI* and the A-weighted speech sound pressure level, respectively, in the horizontal plane situated 1.2 m above the floor of the room with the acoustic treatment taken into account.

Based on the results of numerical simulations, it can be assumed that in order to meet the requirements expressed in terms of the room absorption, the value of the latter should be increased by about 10%–14%. However, because both acoustic prediction and calculation of the room absorption value based on reverberation time values according to Equation (7.2) are only approximations, the next subsection presents results of measurements with the acoustic treatment taken into account and verification of the calculation results obtained with the use of the adopted computational model.

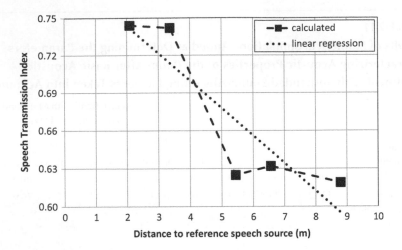

FIGURE 7.20 Results of numerical simulations (carried out in ODEON software) for the room after the basic acoustic treatment with suspended sound-absorbing elements taken into account the speech transmission index *STI* at workstations (the distraction distance r_D is 13.12 m and the privacy distance r_P is 26.76 m).

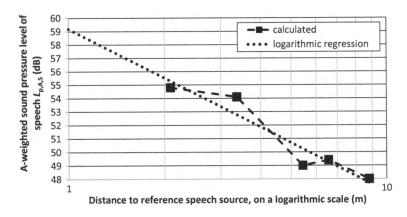

FIGURE 7.21 Results of numerical simulations (carried out with the use of ODEON software) for the room after the basic acoustic treatment with suspended sound-absorbing elements taken into account the A-weighted sound pressure level of speech at workstations (the A-weighted sound pressure level of speech at the distance of 4 m $L_{p,A,S,4\,m}$ is 51.9 dB and the speech spatial decay rate $D_{2,S}$ is 3.62 dB).

7.6.6 MEASUREMENTS OF ACOUSTIC PROPERTIES OF THE ROOM AFTER BASIC ACOUSTIC TREATMENT WITH THE USE OF SUSPENDED SOUND-ABSORBING ELEMENTS: COMPARISON OF RESULTS OF MEASUREMENTS AND NUMERICAL SIMULATIONS

Parameters characterizing acoustic properties of the examined room after basic acoustic treatment with the use of suspended sound-absorbing elements were the subject of empirical studies.

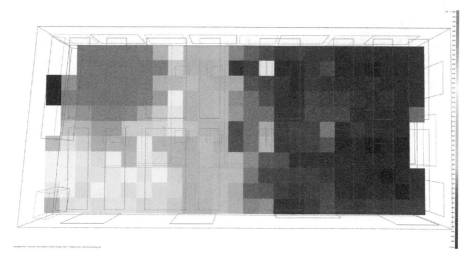

FIGURE 7.22   The speech transmission index value distribution pattern on horizontal plane ($h = 1.2$ m) for the examined room in which the basic acoustic treatment with suspended sound-absorbing elements was taken into account (numerical simulation results by ODEON software).

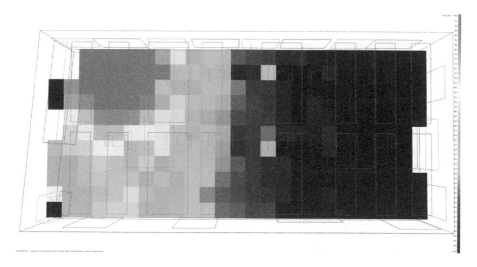

FIGURE 7.23   The A-weighted sound pressure level of speech distribution pattern (denoted SPL(A)) on horizontal plane ($h = 1.2$ m) for the examined room in which the basic acoustic treatment with suspended sound-absorbing elements was taken into account (numerical simulation results by ODEON software).

Results of measurements and numerical simulations for the room after basic acoustic treatment with suspended sound-absorbing elements are summarized in Table 7.12.

The room absorption (calculated from results of reverberation time measurements with the use of Equation 7.2), standardized to 1 m² of orthographic projection of the

TABLE 7.12

Results of Numerical Simulations and Measurement Aimed at Determination of Parameters Characterizing Acoustic Properties of the Room with Basic Acoustic Treatment, with Suspended Sound-Absorbing Elements Taken into Account

| | | | | | In Octave Band with Mid-Band Frequency of (kHz) | | | | | | | | |
| | | | | | 0.5 | 1 | 2 | 0.5 | 1 | 2 | 0.5 | 1 | 2 |
Research Method	r_D (m)	r_P (m)	$L_{p,A,S,4\,m}$ (dB)	$D_{2,S}$ (dB)	T (s)			A (m²)			$A_{1\,m^2}$ (m²)		
Simulation	13.12	26.76	51.9	3.62	0.49	0.5	0.51	72.4	70.9	69.5	0.99	0.97	0.95
Measurement	12.3	22.9	49.5	4.3	0.34	0.41	0.46	104.3	86.5	77.1	1.42	1.18	1.05

Notes:

r_D (m) distraction distance

r_P (m) privacy distance

$L_{p,A,S,4\,m}$ (dB) A-weighted sound pressure level of speech at the distance of 4 m

$D_{2,S}$ (dB) speech spatial decay rate

T (s) reverberation time of the room

A (m²) the room absorption (calculated from the reverberation time T)

$A_{1\,m^2}$ (m²) the room absorption standardized to 1 m² of orthographic projection (surface area) of the room (m²)

light gray shading: the criterion is met

dark gray shading: the criterion is not met

room for frequencies 0.5 kHz, 1 kHz, and 2 kHz, is 1.42 m², 1.18 m², and 1.05 m², respectively. Therefore the parameter meets the basic requirements in octave frequency bands centered at 0.5 kHz and 1 kHz, but is too small by about 3.6 m² (4.5%) for the 2 kHz octave band. That is not, however, a direct consequence of longer reverberation time for the frequency 1 kHz before acoustic treatment, as it has been already noted earlier (cf. Section 7.6.3).

Relative differences of values of parameters characterizing acoustic properties of the examined room obtained from measurements and numerical simulations after basic acoustic treatment with the use of suspended sound-absorbing elements are presented in Table 7.13.

The relative difference between the room absorption values as calculated and measured is included in the range from −9.9% to −30.6%. Relative differences for the remaining three parameters (according to additional requirements of EN ISO 3382-3:2012) are included in the range from −15.8% to +6.7%. That means significant discrepancies between results of numerical modeling and *in situ* measurements. However, they can be still considered acceptable from the point of view of needs typical for the acoustic design development process. Such a liberal approach is possible in view of the fact that the calculated values are biased, for all acoustic parameters, toward the direction safe from the point of view of the acoustic design work.

TABLE 7.13

Relative Differences Between Values of Parameters Characterizing Acoustic Properties of the Examined Room Obtained from Measurements and Numerical Simulations, After Basic Acoustic Treatment with Suspended Sound-Absorbing Elements

			In Octave Band with Mid-Band Frequency of (kHz)		
			0.5	1	2
δr_D (%)	$\delta L_{p,A,S,4\,m}$ (%)	$\delta D_{2,S}$ (%)	Relative Difference in Room Absorption δA (%)		
6.7	4.8	−15.8	−30.6	−18.0	−9.9

Notes:

δr_D (%) relative difference between results of measurements and numerical simulations for the distraction distance

$\delta L_{p,A,S,4\,m}$ (%) relative difference between results of measurements and numerical simulations for the A-weighted sound pressure level of speech at the distance of 4 m

$\delta D_{2,S}$ (%) relative difference between results of measurements and numerical simulations for the speech spatial decay rate

δA (%) relative difference between results of measurements and numerical simulations for the room absorption

Actually, the parameters on which upper limits are imposed (the distraction distance and the A-weighted sound pressure level of speech at the distance of 4 m) obtained from numerical simulations have values higher than those obtained from measurements. On the other hand, calculated values of the parameters on which lower limits are imposed (the room absorption and the speech spatial decay rate) are less than those obtained from measurements. It follows from the above that when the design is based on numerical simulation methods, the acoustic effect happens to be underestimated which should be considered a good design practice.

To sum up it can be concluded that when the simplest approach to calculation of the resultant room absorption is used consisting in arithmetic summation of sound absorption values for individual elements, the obtained room absorption value was underestimated by about 4%, whereas as a result of numerical simulations, all the key room acoustic parameters are underestimated compared to those obtained from measurements.

In view of the above it is possible to accept the simple computational method set out in PN-B-02151-4:2015 based on the room absorption (increasing its value, for safety reasons, by a margin of about 5%) as a reliable tool suitable for acoustic design projects aimed at meeting the basic requirements set out in PN-B-02151-4:2015 (i.e., arriving at the room absorption value not less than 1.1 m² per 1 m² of the room surface area). There is also an option to use a computational simulation method based on a model developed in ODEON software as a reliable tool for design and acoustic prediction in an open-plan office room aimed at meeting basic requirements according to PN-B-02151-4:2015 and additional requirements for the so-called "good acoustic conditions" as per EN ISO 3382-3.

7.6.7 Computational Analysis of the Effect of Additional Acoustic Treatment Elements (Additional Treatment) on Meeting Additional Requirements for Open-Plan Office Rooms as per EN ISO 3382-3

The method used in Sections 7.6.4 through 7.6.6 was the numerical simulation based on the room model implemented in ODEON software. The obtained results allowed the statement that the method can be employed in design work on acoustic treatment of rooms and the observed divergence between results of calculations and measurements show a trend safe from the point of view of the desired acoustical effect (numerical simulation results underestimate the actual effect of acoustic treatment).

In the following, numerical simulation studies are presented for the same room with basic acoustic treatment (with suspended sound-absorbing elements) and with such additional acoustic treatment elements which enable the room to meet both the basic requirements of PN-B-02151-4:2015 and the additional requirements of EN ISO 3382-3:2012.

In the room with basic acoustic treatment described in Sections 7.6.2 and 7.6.3 (20 sound-absorbing elements suspended under the ceiling made of a material with the weighted sound absorption coefficient α_w of 0.6), the following additional acoustic treatment elements were introduced:

- replacement of the sound-absorbing material of suspended sound-absorbing elements with a material with the weighted sound absorption coefficient $\alpha_w = 0.9$,
- installation of three desktop screens, each 1-meter high, placed between workstations, with their surfaces characterized with the weighted sound absorption coefficient $\alpha_w = 0.9$,
- installation of three 1.7-m high acoustic free-standing floor screens, with their surfaces characterized with the weighted sound absorption coefficient $\alpha_w = 0.9$,
- addition of four sound-absorbing wall elements (with dimensions 1.2 m × 1.4 m each) mounted on walls between windows, with the weighted sound absorption coefficient $\alpha_w = 0.9$, and
- use of sound-absorbing material with the weighted sound absorption coefficient $\alpha_w = 0.9$ on the whole ceiling surface.

The room with both basic and additional acoustic treatment taken into account is shown in Figure 7.10, and its schematic view is shown in Figure 7.24.

Results of calculation of parameters determining acoustic properties of the examined room with basic and additional acoustic treatment taken into account are quoted in Table 7.14 and presented in Figures 7.25 and 7.26.

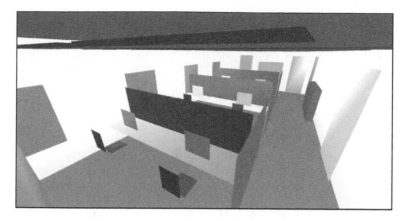

FIGURE 7.24 A schematic view of the room with basic and additional acoustic treatment taken into account (generated in ODEON software).

TABLE 7.14

Results of Numerical Simulations Aimed at Determination of Parameters Characterizing Acoustic Properties of a Room with Basic and Additional Acoustic Treatment Taken into Account

				In Octave Band with Mid-Band Frequency of (kHz)					
				0.5	1	2	0.5	1	2
Required/Calculated	r_D (m)	$L_{p,A,S,4m}$ (dB)	$D_{2,S}$ (dB)	Reverberation Time (s)			Room Absorption (m²)		
Basic requirements as per PN-B-02151-4:2015	—	—	—				≥80.74		
Additional requirements as per EN ISO 3382-3:2012 for "good acoustic conditions"	≤5	≤48	≥7						
Calculated	5.0	41.3	8.4	0.38	0.31	0.39	93.3	114.4	90.9

Notes:

r_D (m) distraction distance

$L_{p,A,S,4m}$ (dB) A-weighted sound pressure level of speech at the distance of 4 m

$D_{2,S}$ (dB) speech spatial decay rate

T (s) reverberation time (for frequencies 0.5 kHz, 1 kHz, and 2 kHz)

A (m²) room absorption (calculated from the reverberation time value for frequencies 0.5 kHz, 1 kHz, and 2 kHz)

light gray shading: the value meets the requirement

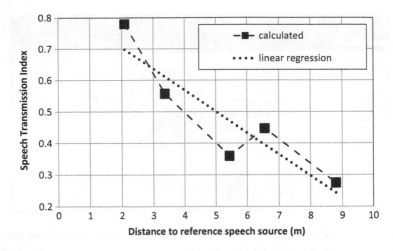

FIGURE 7.25 Results of numerical simulations for the room with basic and additional acoustic treatment taken into account the speech transmission index *STI* at workstations (the distraction distance r_D is 5.0 m, and the privacy distance r_P is 9.45 m).

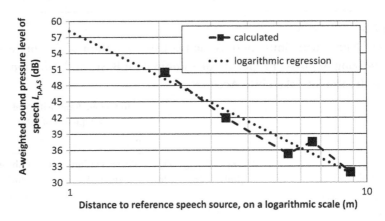

FIGURE 7.26 Results of numerical simulations for the room with basic and additional acoustic treatment taken into account: the A-weighted sound pressure level of speech at workstations (the A-weighted sound pressure level of speech at the distance of 4 m $L_{p,A,S,4\,m}$ is 41.3 dB, and the speech spatial decay rate $D_{2,S}$ is 8.4 m).

The next two figures depict results of numerical simulations for the room with basic and additional acoustic treatments taken into account, of which Figure 7.27 shows the distribution pattern of values of the speech transmission index *STI*, and Figure 7.28—the distribution pattern of the A-weighted sound pressure level of speech. It follows from the form of the distribution patterns that at the height of a seated worker's head ($h = 1.2$ m), the speech intelligibility decreased significantly

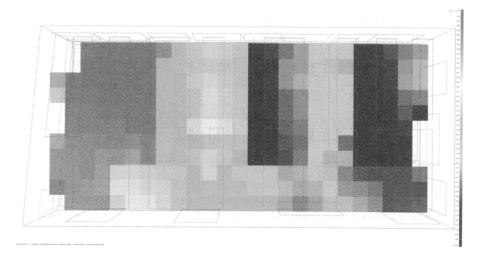

FIGURE 7.27 The distribution pattern of values of the speech transmission index *STI* in a horizontal plane at $h = 1.2$ m above the floor in the room with both basic and additional acoustic treatment taken into account (numerical simulation results by ODEON software).

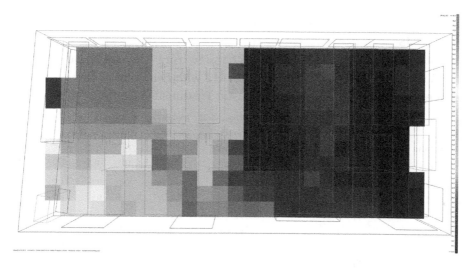

FIGURE 7.28 The distribution pattern of values of the A-weighted sound pressure level of speech (denoted SPL(A)) in a horizontal plane at $h = 1.2$ m above the floor in the room with both basic and additional acoustic treatment taken into account (numerical simulation results by ODEON software).

which is the mostly advisable feature of the room (i.e., the value of the speech transmission index *STI* at the half-length of the room is lower than good, $STI < 0.6$, and drops to about 0.3 at the room end, whereas the A-weighted sound pressure level of speech is lower than the A-weighted sound pressure level of background noise (39 dB) already at half of the room length).

Based on results of numerical simulations it can be concluded that the acoustic treatment configuration adopted in the present chapter enables both the basic requirements according to PN-B-02151-4:2015 and the additional requirements (for "good acoustic conditions") according to EN ISO 3382-3:2012 to be met.

7.6.8 Computational Analysis of the Effect of Modification of Certain Elements of Basic and Additional Acoustic Treatment on Acoustic Properties of the Room

In the next section, an analysis of the possibility to skip one of elements of acoustic treatment (basic and additional) on acoustic properties of the examined room according to additional requirements (EN ISO 3382-3:2012) is carried out.

Results of numerical simulation presented in Table 7.15 indicate that all elements of acoustic treatment suggested to be taken into account in the acoustic treatment design (basic and additional, cf. Section 7.6.7) are necessary to meet the additional requirements concerning acoustic properties of open-plan office rooms according to EN ISO 3382-3:2012.

TABLE 7.15

Measurement Results for Parameters Characterizing Acoustic Properties of the Room According to EN ISO 3382-3:2012 with Omission of One of Elements of Basic and Additional Acoustic Treatment Taken into Account

Sound-Absorbing Ceiling	Suspended Ceiling Sound-Absorbing Elements (with Weighted Sound Absorption Coefficient of)	Wall-Mounted Sound-Absorbing Elements	Height of Acoustic Screens (m)	r_D (m)	r_P (m)	$L_{p,A,S,4\,m}$ (dB)	$D_{2,S}$ (dB)
Applied	Applied (0.9)	Applied	1.7	5.0	9.5	41.3	8.4
Applied	Applied (0.6)	Applied	1.7	5.2	9.8	41.7	8.2
—	Applied (0.9)	Applied	1.7	7.0	12.5	46.1	6.9
Applied	—	Applied	1.7	5.9	11.7	42.9	7.0
Applied	Applied (0.9)	—	1.7	5.2	9.9	41.9	8.0
Applied	Applied (0.9)	Applied	1.6	5.4	10.7	42.1	7.6

Notes:

r_D (m) distraction distance

r_P (m) privacy distance

$L_{p,A,S,4\,m}$ (dB) A-weighted sound pressure level of speech at the distance of 4 m

$D_{2,S}$ (dB) speech spatial decay rate

light gray shading: the criterion is met

dark gray shading: the criterion is not met

7.7 RECOMMENDATIONS CONCERNING NECESSARY ACOUSTIC TREATMENT OF OPEN-PLAN OFFICE ROOMS

Today's open-plan office rooms are strongly diversified as far as their overall size and the number of workstations is concerned, and more than anything else, with regard to interior arrangement and acoustic treatment. From the point of view of acoustic properties of open-plan office rooms, the following acoustic treatment elements should be considered necessary:

- acoustic separation of the open-plan office room from other spaces serving other purposes, and in particular: canteens, cafeterias, kitchens, welfare and rest areas, meeting and conference rooms (including areas dedicated for making/receiving individual telephone calls), rooms for confidential talks, and shared office equipment (photocopiers, plotters, servers, switchboxes, etc.);
- in open-plan office rooms, separation (at least with high acoustic screens [EN ISO 17624:2004]) of traffic routes from workstations;
- covering the floor with fitted carpet as thick as practically possible and with the highest possible sound absorption coefficient;
- installation of suspended sound-absorbing ceilings (as distant as possible from the solid ceiling structure) with the sound absorption coefficient 0.9 (ceilings with lower sound absorption coefficient on only a portion of the ceiling surface may be insufficient to meet all the criteria discussed in Section 7.3);
- mounting of sound-absorbing material panels on room walls (in quantities and with acoustic properties of materials adequate to meet all the criteria specified in Section 7.3);
- setting out acoustic screens (1.5–1.7 m high) on the room floor covered with a sound-absorbing material (effective enough to meet all the criteria specified in Section 7.3); and
- consideration of technical and organizational (worker's acceptance) feasibility of using speech-masking sound sources (with the impassable threshold of 35–40 dB for the A-weighted sound pressure level of such sources produced at all workstations).

7.8 SUMMARY

Empirical studies carried out by the present author in 15 open-plan office rooms confirm the argument that the majority of rooms currently used for that purpose fail to meet the acoustic criteria determined in the international standard EN ISO 3382-3:2012 and in Polish standard PN-B-02151-4:2015. The basic requirements determined in PN-B-02151-4:2015 (concerning room absorption) are met in about 27% of the examined cases in all the three frequency bands covered in the analysis (0.5 kHz, 1 kHz, and

2 kHz); whereas in about 27% of cases, the basic requirements are met in neither of the above-mentioned frequency bands. In the remaining rooms, the room absorption value is sufficient in one or two frequency bands. The additional requirements for "good acoustic conditions" established in EN ISO 3382-3:2012 (based on three parameters: the distraction distance r_D, the A-weighted sound pressure level of speech at the distance of 4 m, and the speech spatial decay rate) are not met by any of the rooms. About 33% of the rooms meet the additional requirements only for some of the criteria parameters (especially A-weighted sound pressure level of speech at the distance of 4 m). The main reason why the currently operated rooms fail to meet the requirements (according to PN-B-02151-4:2015 and/or EN ISO 3382-3:2012) is the fact that the requirements were established in recent years (2012 and 2015, respectively). Since they are not yet legally binding, they were not taken into account in the period where the rooms were designed and constructed. This proves that acoustic conditions prevailing in these premises are not appropriately adapted to the nature of the work carried out in them. As there is no direct hazard of hearing damage to workers and the A-weighted sound pressure level adopted as the arduousness threshold (which, for instance, is 55 dB according to Polish standard PN-N-01307:1994) is not exceeded either, the noise levels cannot be formally considered excessive. However, characteristic features of the noise (the sound waveform pattern, spectrum, reverberative nature), result in excessive intelligibility of third-parties' conversations which is a work-disrupting factor. In view of the above, precedence must be given to the necessity to adapt acoustic properties of the examined rooms to applicable acoustic standards by developing and embodying an appropriate acoustic treatment design. Based on relevant measurements carried out in the analyzed rooms it can be claimed that the necessary measures (which, however, cannot be considered sufficient for all of the rooms) included installation of: suspended sound-absorbing ceilings with the sound absorption coefficient of the order 0.9; sound-absorbing carper flooring; sound-absorbing materials on walls; tall acoustic screens between workstations lined on both sides with sound-absorbing materials; and maintaining (with the use of masking sources, where necessary) the A-weighted sound pressure level of background noise on the level of about 35–40 dB. The studies prove that in the design development process, it is necessary to approach each examined room individually and appropriately, take into account various solutions, and verify them numerically with the use of dedicated software for sound field simulation in rooms. In newly designed open-plan office rooms, it is also necessary to remove all unnecessary sources of noise and ensure insulation from noise penetrating into the room from outside. A good idea to consider in this respect would be to section off separate areas for photocopiers, printers, plotters, welfare amenities, and rooms dedicated to making and receiving phone calls.

ACKNOWLEDGMENTS

The author is deeply indebted to the team of his colleagues from the Central Institute For Labour Protection—National Research Institute for their assistance in taking measurements in the years 2014–2019 [Mikulski et al. 2016, 2019], especially those shown in Tables 7.5 and 7.7.

REFERENCES

Asselineau, M., and A. Gaulupeau. 2017. Trouble shooting in an open plan office: Case study. *Proceeding of the 24th International Congress on Sound and Vibration.* July 23–27, 2017, London, Great Britain.

Bradley, J. S., R. Reich, and S. G. Norcroos. 1999. A just noticeable difference in C50 for speech. *Appl Acoust* 58(2):98–108. DOI:10.1016/S0003-682X(98)00075-9.

Cabrera, D., M. Yadav, and D. Protheroe. 2018. Critical methodological assessment of the distraction distance used for evaluating room acoustic quality of open-plan offices. *Appl Acoust* 140:132–142. DOI:10.1016/j.apacoust.2018.05.016.

Davidsson, F., and P. Hodsman. 2017. Speech propagation in open-plan office: A cross over designed field study. *Proceeding of the 24th International Congress on Sound and Vibration.* July 23–27, 2017, London, Great Britain.

Ebissou, A., E. Parizet, and P. Chevret. 2015. Use of the speech transmission index for the assessment of sound annoyance in open-plan offices. *Appl Acoust* 88:90–95. DOI:10.1016/j.apacoust.2014.07.012.

Edgington, C., and M. Stevens. 2017. Practical considerations and experiences with sound masking's latest technology. *Proceeding of the 24th International Congress on Sound and Vibration.* July 23–27, 2017, London, Great Britain.

EN ISO 3382-2:2008. 2008. Acoustics—Measurement of room acoustic parameters—Part 2: Reverberation time in ordinary rooms. Brussels, Belgium: European Committee for Standardization.

EN ISO 3382-3:2012. 2012. Acoustics—Measurement of room acoustic parameters—Part 3: Open plan offices. Brussels, Belgium: European Committee for Standardization.

EN ISO 9921:2003. 2003. Ergonomics—Assessment of speech communication. Brussels, Belgium: European Committee for Standardization.

EN ISO 11201:2010. 2010. Acoustics—Noise emitted by machinery and equipment–Determination of emission sound pressure levels at a work station and at other specified positions in an essentially free field over a reflecting plane with negligible environmental corrections. Brussels, Belgium: European Committee for Standardization.

EN ISO 17624:2004. 2004. Acoustics—Guidelines for noise control in offices and workrooms by means of acoustical screens. Brussels, Belgium: European Committee for Standardization

EN 60268-16:2011. 2011. Sound system equipment—Part 16: Objective rating of speech intelligibility by speech transmission index. Brussels, Belgium: European Committee for Standardization.

Evans, G. W., and D. Johnson. 2000. Stress and open-office noise. *J Appl Psychol* 85(5):779–783. DOI:10.1037/0021-9010.85.5.779.

Haapakangas, A., and V. Hongisto. 2017. Distraction distance and perceived disturbance by noise–An analysis of 21 open-plan offices. *J Acoust Soc Am* 141(1):127–136. DOI:10.1121/1.4973690.

Haapakangas, A., V. Hongisto, J. Hyönä, J. Kokko, and J. Keränen. 2014. Effects of unattended speech on performance and subjective distraction: The role of acoustic design in open-plan offices. *Appl Acoust* 86:1–16. DOI:10.1016/j.apacoust.2014.04.018.

Keränen, J., and V. Hongisto. 2013. Prediction of the spatial decay of speech in open-plan offices. *Appl Acoust* 74(12):1315–1325. DOI:10.1016/j.apacoust.2013.05.011.

Kim, J., and R. de Dear. 2013. Workspace satisfaction: The privacy-communication trade-off in open-plan offices. *J Environ Psychol* 36:18–26. DOI:10.1016/j.jenvp.2013.06.007.

Kostallari, K., E. Parizet, P. Chevret, J.-N. Amato, and E. Galy. 2017. Irrelevant speech effect in open plan offices: A laboratory study. *Proceeding of the 24th International Congress on Sound and Vibration.* July 23–27, 2017, London, Great Britain.

Kulowski, A. 2011. Akustyka Sal. Zalecenia projektowe dla architektów. [Room acoustics. Design recommendations for architects]. Publisher: Gdańsk University of Technology. https://www.ksiegarniatechniczna.com.pl/akustyka-sal-zalecenia-projektowe-dla-architektow.html.

Mikulski, W., I. Warmiak, D. Pleban, J. Kozłowski, and D. Bielecka. 2016. Badania propagacji dźwięku i metod kształtowania warunków akustycznych w pomieszczeniach do pracy wymagającej koncentracji uwagi. [Research on sound propagation and shaping acoustic conditions in rooms designed for work requiring concentration of the mind]. Warszawa: CIOP-PIB.

Mikulski, W. 2016. Warunki akustyczne w pomieszczeniach biurowych open space–wyniki badań pilotażowych. [Acoustic conditions in open plan offices—Pilot test results.]. *Med Pr* 67(5):653–662. DOI:10.13075/mp.5893.00425.

Mikulski, W., J. Radosz, D. Pleban et al. 2019. Badania oraz opracowanie metody kształtowania akustycznego środowiska pracy w wielkoprzestrzennych pomieszczeniach do pracy umysłowej. [A research work plan and development of a method for shaping acoustic work environment in large-space rooms designed for intellectual work]. Warszawa: CIOP-PIB.

Mikulski, W. 2018. Warunki akustyczne w pomieszczeniach biurowych open space–zastosowanie środków technicznych w typowym pomieszczeniu. [Acoustic conditions in open plan offices—Application of technical measures in a typical room]. *Med Pr* 69(2):153–165. DOI:10.13075/mp.5893.00574.

Mikulski, W. 2019a. Badania obliczeniowe zrozumiałości mowy w pomieszczeniach biurowych open space. [Computational studies of speech intelligibility in open-plan offices]. *Med Pr* 70(3):327–342. DOI:10.13075/mp.5893.00726.

Mikulski, W. 2019b. Projektowanie adaptacji akustycznej otwartych pomieszczeń do prac administracyjnych. Część 1. Projektowanie podstawowe. [Acoustic adaptation designing of open space for administrative works. Part 1. Basic design]. *Materiały Budowlane* 8:16–20. DOI:10.15199/33.2019.08.02.

Muller-Trapet, M., and B. N. Gover. 2019. Relationship between the privacy index and ite speech privacy class. *J Acoust Soc Am* 145(5):EL435–EL441. DOI:10.1121/1.5109049.

Nurzyński, J. 2018. Warunki akustyczne w wielkoprzestrzennych pomieszczeniach biurowych. [Acoustic conditions in open space offices]. *Materiały Budowlane* 552(8):10–12. DOI:10.15199/33.2018.08.02.

ODEON Computer software. https://odeon.dk/. (accessed January 28, 2020).

PN-N-01307:1994. 1994. Hałas—Dopuszczalne wartości parametrów hałasu w środowisku pracy—Wymagania dotyczące wykonywania pomiarów [Polish standard: Noise—Permissible values of noise parameters in the work environment—Requirements for performing measurements]. Warsaw, Poland: Polski Komitet Normalizacyjny [Polish Committee for Standardization].

PN-B-02151-2:2018. 2018. Akustyka budowlana. Ochrona przed hałasem pomieszczeń w budynkach. Wymagania dotyczące opuszczalnego poziomu dźwięku w pomieszczeniach [Polish standard: Building acoustics. Noise protection of rooms in buildings. Requirements for the permissible sound pressure level in rooms]. Warsaw, Poland: Polski Komitet Normalizacyjny [Polish Committee for Standardization].

PN-B-02151-4:2015. Akustyka budowlana. Ochrona przed hałasem w budynkach Część 4: Wymagania dotyczące warunków pogłosowych i zrozumiałości mowy w pomieszczeniach [Polish standard: Building acoustics. Noise protection in buildings Part 4: Requirements for reverberation time and speech transmission index in rooms]. Warsaw, Poland: Polski Komitet Normalizacyjny [Polish Committee for Standardization].

Renz, T., P. Leistner, and A. Liebl. 2018. The effect of spatial separation of sound masking and distracting speech sound on working memory performance and annoyance. *Acta Acust United Acust* 104(4):611–522. DOI:10.3813/AAA.919201.

Rindel, J. H. 2012. Prediction of acoustical parameters for open plan offices according to ISO 3382-3. *J Acoust Soc Am* 131(2012):3357–3357. DOI:10.1121/1.4708587.

Rindel, J. H., and C. L. Christensen. 2012. Acoustical simulation of open-plan offices according to ISO 3382-3, *Proceeding of the Conference Euronoise 2012*. June 10–13, 2012, Prague, Czech Republic.

Smith, V. 2017. Using acoustical modeling software to predict speech privacy in open-plan offices. *J Acoust Soc Am* 141(5):3598–3598. DOI:10.1121/1.4987691.

Vellenga-Persoon, S., T. Hongens, and T. Bouwhuis. 2017. Proposed method for measuring liveliness in open plan offices. *Proceeding of the 24th International Congress on Sound and Vibration*. July 23–27, 2017, London, Great Britain.

Virjonen, P., J. Keränen, and V. Hongisto. 2009. Determination of acoustical conditions in open-plan offices: Proposal for new measurement method and target values. *Acta Acust United Acust* 95:279–290. DOI:10.3813/AAA.918150.

Yadav, M., J. Kim, D. Cabrera, and R. de Dear. 2017. Auditory distraction in open-plan office environments: The effect of multi-talker acoustics. *Appl Acoust* 126:68–80. DOI:10.1016/j.apacoust.2017.05.011.

Yu, J. X., S. P. Wang, X. J. Qiu, A. Shaid, and L. J. Wang. 2016. Contributions of various transmission paths to speech privacy of open ceiling meeting rooms in open-plan offices. *Appl Acoust* 112:59–69. DOI:10.1016/j.apacoust.2016.05.002.

Zaglauer, M., H. Drotleff, and A. Liebl. 2017. Background babble in open-plan offices: A natural masker of disruptive speech? *Appl Acoust* 118:1–7. DOI:10.1016/j.apacoust.2016.11.004.

8 Ultrasonic Noise Measurements in the Work Environment

Jan Radosz

CONTENTS

8.1 INTRODUCTION

Criteria concerning occupational exposure to ultrasounds applicable in different countries cover the range of frequencies from 6.3 kHz to as much as 125 kHz [Lawton 2013]. It can be therefore acknowledge that the ultrasonic noise is a (undesirable) sound the spectrum of which contains components with high audible and low ultrasonic frequencies. The main sources of ultrasonic noise in the work environments are the so-called low-frequency ultrasonic technological process devices such as washers, welders, erosion machines, handheld soldering guns, and tin plating tanks [Pawlaczyk-Łuszczyńska et al. 2007; Mikulski and Radosz 2010]. Ultrasonic noise is also emitted by high-speed compressors, torches, valves, pneumatic tools, and high-speed machine tools such as planers, milling machines, circular saws, and some textile industry machines [Smagowska 2013b]. Currently, recommended measuring methods do not include any information concerning accuracy of measurements or estimation of the uncertainty budget for ultrasonic noise. Literature of the subject lacks information about factors affecting the sound pressure level measurement results in the frequency range specific for ultrasonic noise (especially above 20 kHz), whereas analysis of these factors is necessary in order to increase precision and reliability of the related risk assessment. Further, no currently applicable international standards exist concerning measurements of ultrasonic noise in the work environment [Radosz and Pleban 2018]. Therefore, there is an urgent need to standardize the measuring methods developed over the space of last two decades, including review of the requirements concerning determination of the measurement uncertainty budget and current availability of measuring apparatus options and calibration techniques.

8.2 A REVIEW OF REGULATIONS APPLICABLE IN POLAND AND OTHER COUNTRIES TO ULTRASONIC NOISE AT WORKSTATIONS

Selected criteria of occupational exposure to ultrasonic noise are listed in Table 8.1.

8.3 THE ULTRASONIC NOISE EFFECT ON HUMAN BODY

Many research results indicate that beyond a shadow of a doubt, airborne ultrasounds can penetrate into the human body via the hearing organ and through the whole surface of the body despite absence of any specific receptors in the skin, in a way similar to electromagnetic or ionizing radiation. In view of this fact, efforts must be made to distinguish between these two types of ultrasonic noise effects on humans: the effect on the hearing organ (the ear) and the so-called extra-aural effects [Koton 1986; Acton 1973, 1974].

One difficulty encountered in research on the effect of ultrasonic noise on condition of the human ear (hearing organ) involves the fact that under industrial conditions, ultrasounds are usually accompanied by audible noise and it is hard to determine whether hearing issues in subjects occur as a result of exposure to either audible components or ultrasonic components alone or if they are rather due to simultaneous exposure to both of the two noise components [Śliwiński 2001, 2016]. However,

TABLE 8.1
Criteria for Assessment of Occupational Exposure to Ultrasonic Noise Applicable in Selected Countries

Country/ Organization/ Author, Year of Issuance	Third-Octave Band Mid-Band Frequency (kHz)													
	6.3	8	10	12.5	16	20	25	31.5	40	50	63	80	100	125
	Maximum Admissible Sound Pressure Level in Third-Octave Frequency Bands (dB)													
Japan, 1971[a]	90	90	90	90	90	110	110	110	110	110				
USSR, 1989[b]				80	80(90)	100	105	110	110	110	110	110	110	
Acton, 1975[c]				75	75	75	110	110						
USA, 2001[d]			80	80	80	105	110	115	115	115				
England, 1977[e]				75						110				
Norway, 1978[f]								120			120			120
Bulgaria, 1979[g]				75	85	110	110	110	110	110	110	110	110	
Canada, 1980[h]	80	80	80	80	80	80	110	110	110	110	110	110		
Australia, 1981[i]		75	75	75	75	110	110	110	110					
IRPA, 1981[j]		80	80	80	80	80	110	110	110	110	110	110	110	
IRPA, 1984[k]					75	110	110	110	110	110	110	110	110	
Sweden, 1992[l]						105	115	115	115	115	115	115	115	115
Poland, 2018[m]			80	80	80	90	105	110	110					

[a] A Circular of the Japanese Ministry of Labour, 1971.
[b] Standard GOST-12.1.001-89 The ultrasound. General safety requirements.
[c] Acton W.I., Exposure criteria for industrial ultrasound. *Ann. Occup. Hyg.*, 1975, 18, 267–268.
[d] American Conference of Governmental Industrial Hygienists 2001.
[e] Sanitary regulations 1977.
[f] Recommendations Utkast til Forskrfter om Stoj pa Arbeidsplassen, 1978 (assessment in octave bands).
[g] Standard BDS-12.1.00.1-79.
[h] Ordinance of the Minister of Health and Social Issues 1980.
[i] Standrd AS2243-1981.
[j] Preliminary recommendations of the International Radiation Protection Association, 1981.
[k] Preliminary recommendations of the International Radiation Protection Association, 1984.
[l] Recommendations of the National Labour and Health Protection Bureau in Sweden (AFS 992:10), 1992.
[m] Recommendation of the Minister of Family, Labour, and Social Policy of 12 June 2018 on maximum admissible concentration and intensities of agents hazardous to health in the work environment.

popularity is growing for the view that due to nonlincar phenomena occurring in the ear alone as a result of excitation with ultrasonic waves, sub-harmonic components may occur with sound pressure levels of sometimes the same order of magnitude as the basic ultrasonic component. As a consequence of the phenomenon, hearing losses are considered to be due just to sub-harmonic frequencies of ultrasonic fields.

Hearing losses in persons occupationally exposed to noise can be accelerated by the effect of simultaneous exposure to ototoxic factors (organic solvents, asphyxiant substances, and heavy metals). During operation of certain ultrasonic machines or devices (e.g., in the course of ultrasonic welding), the nature of the generated sound is close to that of the pulsed noise which may have a significant effect on the extent of hearing damage. An extensive research study on hearing losses in subjects exposed to audible industrial noise with levels above 85 dB in the high-frequency range (above 8 kHz) for a period of eight years revealed that hearing damage thresholds observed in the high-frequency audiometry range (i.e., 8–20 kHz) increased earlier and faster in the examined group of subjects.

Apart from the harmful effect on hearing, there are also reports evidencing negative effects of ultrasounds on the vestibule organ in the inner ear, manifesting themselves in headache and dizziness, disorders of the sense of balance, nausea, drowsiness during the daytime, excessive tiredness, etc. [Smagowska 2013a; Smagowska and Pawlaczyk-Łuszczyńska, 2013].

In the scope of extra-aural effects it turned out that occupational exposure to ultrasonic noise with levels above 80 dB in the range of audible frequencies and above 100 dB in the range of low ultrasonic frequencies resulted in changes of vegetative-vascular nature.

Studies on extra-aural effects revealed that ultrasonic noise is the cause of disorders in the cardiovascular system [Smagowska and Pawlaczyk-Łuszczyńska 2013], manifesting in impaired blood supply to both the cardiac muscle and peripheral tissues. Symptoms of the related disorders include decrease of skin warmth, skin on face and neck turning suddenly pale or red, distinct slowdown of cardiac action (pulse frequency decrease), arterial blood pressure drop, etc. Disorders in the blood circulation system are accompanied by changes in blood composition. Symptoms can include decrease in the number of red blood cells at correct hemoglobin levels which leads to the condition referred to as hyperchromatism.

Ultrasonic noise can also be the cause of thermoregulatory process disorders. Possible symptoms include increase in body temperature by $0.5°C$–$0.7°C$; in these cases the increase is greater when there are higher pressure levels of airborne ultrasounds.

There is also conclusive evidence in support of the adverse effect of ultrasonic noise on metabolic processes and nervous system functions. In operators of ultrasonic devices performing their duties for longer periods, increased neural excitability and the feeling of permanent annoyance can be observed. One frequently appearing condition is high variability of mood, manifesting itself with quick transitions from flying into a temper to apathy. Other symptoms include difficulties of intellectual nature manifesting themselves in memory impairment, problems with concentration, and impairment of the ability to soak up new knowledge.

Highly sensitivity to low-frequency airborne ultrasound also shows in the human endocrine glands, especially gonads and thyroid, where adverse lesions are observed with different degree of graveness depending on intensity of ultrasonic noise and exposure duration time.

Results of survey studies carried out in the J. Nofer Institute of Occupational Medicine and in the Central Institute for Labour Protection—National Research Institute concerning the effect of exposure to ultrasonic noise at workstations provide evidence of adverse extra-aural effect of the noise on workers [Pawlaczyk-Łuszczyńska et al. 2007; Smagowska and Pawlaczyk-Łuszczyńska 2013]. The respondents complained of tiredness, headache, dizziness, drowsiness, and palpitations. The workers maintained that the ultrasonic noise existing in the work environment made it hard to communicate verbally and prevented them from concentrating. The noise was described by workers as loud, uneven, rough and unpleasant, whizzing and squeaky, annoying, irritating, and disturbing the course of work.

The research carried out in laboratory conditions of the Central Institute for Labour Protection—National Research Institute on the effect of ultrasonic noise on cognitive functions and psychophysical fitness in humans revealed that noise containing low ultrasonic frequencies adversely affects continuity of attention, work output, and concentration of mind [Smagowska 2013a].

The harmful effects of airborne ultrasound on the human body are so diverse and affect so many body organs and systems that ultrasonic noise is now counted among other harmful physical factors existing potentially in work environments. Therefore the ultrasonic noise levels are subject to specifically established and enforced standards defining their values acceptable at workstations from the point of view of occupational health, whereas employers who have devices that are considered sources of ultrasonic noise installed in their factories are obliged to take measurements for the purpose of assessment of occupational risk connected with exposure of employees to that type of noise.

8.4   THE MEASURING APPARATUS

As opposed to sound pressure level measurements in the range of audible frequencies (which are carried out with sound level meters the requirements applicable to which are established, for instance, in standard IEC 61672-1), the apparatus for sound pressure level measurements in the range of ultrasonic frequencies is neither defined nor subjected to specific requirements concerning the usage or calibration in the framework of scheduled metrological inspection.

In view of the internationally adopted ultrasound assessment criteria, a plausible solution would consist in taking measurements with the use of a sound meter/analyzer equipped with a microphone appropriately selected for the examined frequency range and third-octave band-pass filters provided that:

- the frequency characteristic of the meter/analyzer covers frequencies from the examined range;
- the range of mid-band frequencies of band-pass filters integrated in the meter/analyzer covers the frequencies for which acceptable values have been determined; and
- the sound pressure level measuring range and the linearity range of the meter/analyzer are adapted to the sensitivity of the microphone and further, it is suitable to take measurements of the sound pressure level in the range corresponding to its values occurring in actual on-site conditions.

To ensure that measurement results are reliable and their uncertainties are correctly estimated it is necessary to be fully aware of current state of the measuring apparatus. Any metrological inspection including verification of consistency of the measuring apparatus (sound level meter/analyzer and acoustic calibrator) with requirements set out in applicable standards (such as, e.g., IEC 61672-1, IEC 61260-1, and IEC

60942) should be carried out periodically in accredited standardizing laboratories. Any scheduled metrological inspection in the relevant frequency range, carried out at least every two years, should include as the minimum:

- microphone calibration with determination its frequency characteristic in free field;
- calibration of third-octave band-pass filters (attenuation characteristics, errors, and linearity range);
- inherent noise measurements; and
- determination of errors due to linear averaging and exponential averaging.

In the range of frequencies specific to the ultrasonic noise, the measuring microphone is the element of special importance. Taking into account, on one hand, the examined ultrasonic noise frequency as well as the minimum and maximum sound pressure levels observed at workstations, and on the other, electroacoustical properties of microphones available commercially on the market, it must be stated that the optimum choice would be free-field microphones with the rated diameter of 1/4″ [Radosz and Pleban 2014].

The measuring microphone in combination with a meter/analyzer should offer flat frequency characteristic covering frequencies in the whole examined range (±0.5 dB). The microphone should be used, first of all, in configuration without protection grid. The use of a microphone in configuration with a protection grid is also possible provided that its effect on the frequency characteristic of the microphone is known.

8.5   THE MEASURING APPARATUS AND CALIBRATION

8.5.1   MICROPHONE ACCESSORIES AND THEIR EFFECT ON MEASUREMENT RESULTS

In general, ultrasonic noise measurements at workstations do not require the use of any windshields, all-weather protection kits, or correction nose cones to be installed on microphones. However, should such need occur, then the frequency characteristic of the microphone in configuration with such accessory should be known in view of its significant effect on the measurement result—up to 5 dB for 40 kHz [Radosz and Pleban 2018]. Although microphone preamplifiers are characterized with low output impedance and high output current value, the use of long extension cables which constitute a leading load for the preamplifier, may affect the output voltage value and the frequency characteristic of the microphone-preamplifier system and decrease the upper limit of the usable frequency range. The effect depends on capacitance of the cables as well as on the upper limit of the measured signal dynamics range and is practically negligible for low-level signals and typical two-meter long cables. To estimate the potential effect of cables on measurement results, it is necessary to make use of the manufacturers' data. Some providers of acoustic measuring equipment (e.g., sound level meters) provide corrections for the effect of extension cables on the measuring system frequency characteristic. A good solution is also determination of the frequency characteristic of the whole system including extension cables of various lengths planned to be used in the course of measuring sessions, as part of the system calibration procedure.

The use of cables can facilitate penetration of radio-frequency electromagnetic interference into the measuring setup via both the preamplifier and other system components. The system may be rendered resistant to such noise by using suitable low-pass filters. Electronic instruments immune to such electromagnetic interference and meeting applicable requirements of directives concerning electromagnetic compatibility are identified with the "CE" mark. Modern instruments are typically tested and certified for resistance to electromagnetic interference, at least on the level of the manufacturer's statement. It should be however borne in mind that presence of *any* component insufficiently immune to interference deteriorates the immunity of the whole system.

Results of ultrasonic noise measurements depend significantly on microphone configuration, in particular on absence or presence of the protection grid, as with increasing frequencies, both size and location dimensions of grid mesh holes may have an effect on sound waves with small wavelengths. Installation of a protection grid may inflate results of measurements by even as much as 5 dB for 40 kHz [Radosz and Pleban 2018].

8.5.2   Factors Pertaining to Adjustment and Calibration of Apparatus with the Use of an Acoustic Calibrator

Calibration of the measuring apparatus with the use of an acoustic calibrator carried out on site before and after measurement-taking sessions may have an essential effect on the uncertainty of sound pressure level measurements. The calibration consists of checking and adjusting the apparatus using a selected flat frequency characteristic in the way ensuring that the obtained indication corresponds to the rated sound pressure level value of the calibrator in actual environmental conditions. The calibration error effected that way should be 0 dB in the ideal case. In view of the course of the frequency characteristic of a measuring instrument with a 1/4-inch microphone, the pressure-based calibration with the use of an acoustic calibrator with the rated frequency of 1 kHz can be considered equivalent to calibration in the free field conditions. The quantities identified as those affecting the result of calibration with the use of a calibrator are presented in Figure 8.1.

Repeatability of the measuring apparatus' response to the calibrator signal can be affected by external factors such as imprecision of coupling between the calibrator chamber and the microphone and mechanical imperfections of the calibrator-microphone coupling chamber structure. The repeatability depends only to a minor degree on the background noise pressure level in view of the sound-insulating properties of the calibrator coupling chamber and on instability of electroacoustical properties of the measuring apparatus itself which, even if existed, would be masked by relatively low resolution of the meter display.

However, a significant effect on the measurement uncertainty may be due to the properties of the selected acoustic calibrator, in particular:

*   the reference sound pressure level produced by the calibrator and the uncertainty to which the parameter was determined in the course of calibration in a laboratory, holding official accreditation where possible;

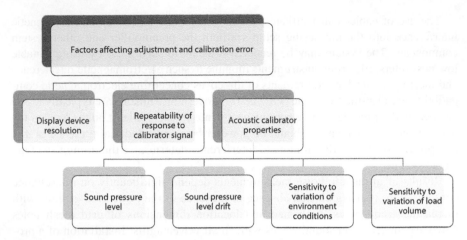

FIGURE 8.1 Factors affecting the adjustment and calibration error of the measuring apparatus with the use of an acoustic calibrator.

- stability of the sound pressure level determined in terms of the drift or the level difference between successive calibrations; and
- sensitivity of the device to variations in static pressure, temperature, and relative humidity changes relative to their reference values established as 101.325 kPa, 23°C, and 50%, respectively, described in terms of sensitivity coefficients determined and declared by the manufacturer.

The scale of effects due to the above-listed factors depends, on one hand, on the quality of the calibrator decisive for stability of the sound level its produces and sensitivity to quantities affecting its value, and on the other, on the length of the intervals between and regularity of successive calibrations.

8.5.3 Factors Connected with the Form of Frequency Characteristics in Free Field

The measured sound pressure level in the third-octave frequency band with the mid-band frequency (mbf) is biased with an error resulting from:

- the non-zero value of the flat frequency characteristic for the measuring apparatus in free field (for an ideal characteristic, the values are 0 dB in the whole frequency range);
- the unevenness of the flat frequency characteristics of the measuring apparatus within individual third-octave frequency bands; and
- the relative attenuation of individual third-octave filters in their pass bands.

Values of the quantities listed above are determined as part of periodical scheduled metrological inspection or estimated on the grounds of representative tests.

8.5.4   Sources of Unevenness and Inaccuracy of Measurement Apparatus Indications

The following are the sources of unevenness and inaccuracy of indications of the measuring apparatus:

a. *The effect of relative attenuation frequency characteristics of filters*—for specific frequencies, a third-octave filter has standardized relative attenuation minimum and maximum values within which attenuation values of actual filter should be included. That means, however, that frequency characteristics of actual filters meeting requirements of applicable standards may differ from each other and have different effect on results of measurements.

b. *The effect of actual relative bandwidth of filters for individual mid-band frequencies*—third-octave band-pass filters should have a constant relative bandwidth which means that the pass band width represents a constant percentage of the mid-band frequency for each band-pass filter. A parameter representing that property is the effective relative filter bandwidth and its error is calculated for an actual filter relative to a reference value.

c. *The effect of linearity errors and level range switchovers*—the measuring apparatus should respond linearly to sound pressure level changes in the whole linearity regime and to each change of measuring level range. Nevertheless, linearity errors must not be used for adjusting results of measurements of real-life complex signals. However, error values should be taken into account in confidence estimations.

d. *The effect of time characteristics relating to signal averaging* (linear or exponential)—accuracy of measurements of the equivalent sound pressure level or the exponentially averaged sound pressure level depends on the capability of the measuring apparatus in scope of averaging which is determined by errors of response to tone pulses or series of tone pulses established for a sound level meter in the range of audible frequencies with the use of the method set out in IEC 61672-3. However, the approach cannot be applied to adjustment of measurement results in case of actual complex signals, especially when carried out in third-octave bands. However, the errors should be taken into account in estimation of confidence limits connected with the measuring apparatus.

e. *The effect of the directivity characteristic*—in the range of ultrasonic frequencies, the phenomenon concerns mainly the directivity pattern of microphones used typically with an extension cable. The effect increases with increasing frequency, as the directivity pattern of a 1/4-inch microphone, approximately omnidirectional in the range of audible frequencies, becomes directional for frequencies higher than 20 kHz. Additionally, the effect depends on microphone configuration and is definitely more distinct for microphones provided with a protection grid. For instance, for microphones type 4135, the manufacturer (Brüel & Kjær) declares omnidirectionality within ±1 dB in the range of frequencies up to 10 kHz and within ±3 dB in the range of up to 20 kHz. For higher frequencies, directivity of the microphone increases significantly.

f. *The effect of inherent noise*—the intrinsic noise associated with and originating from the measuring apparatus in third-octave bands determines the lower limit of the measuring range and may affect results of measurements of low-value sound pressure level. The effect of inherent noise becomes significant when the measurement result level to the noise level ratio drops below 10 dB.

8.5.5 Errors Due to Physical Environmental Conditions

The measuring errors due to effects of physical factors depend, among other things, on sensitivity of the measuring apparatus to variation of static pressure, temperature, and humidity. Environmental conditions affect, first of all, the microphone, but they can be also detrimental to other elements of the measuring apparatus. In the range of audible frequencies, standards determine acceptable variation of indications within predetermined variation ranges of static pressure, temperature, and humidity. Similarly, standards concerning band-pass filters set out admissible limits for variation of filter attenuation in determined ranges of variability of environmental conditions. On the other hand, there are neither requirements nor data concerning the scale of the effect of environmental conditions in case of measuring apparatus equipped with a 1/4-inch microphone when used in the range of ultrasonic noise frequencies.

It is a well-known fact that the microphone response depends on environmental conditions which affect acoustic properties of air trapped between the microphone membrane and the fixed electrode as well as in the cavity behind the electrode. The effect of static pressure, temperature, and humidity, tested and declared by microphone manufacturers is described by means of sensitivity coefficients determining variation of microphone response level due to variation of these quantities relative to the response level in the reference conditions (static pressure of 101.325 kPa, temperature 23°C, relative humidity 50%).

Information concerning dependence of response of working microphones on static pressure, temperature, and humidity expressed as respective functions of frequency are in general hardly available. It should be however underlined that adjustment and calibration of measuring apparatus carried out on-site in actual environmental conditions with the use of an acoustic calibrator with correctly determined and well-known electroacoustic properties, compensates for the effect to some extent provided the measurements are not excessively prolonged and the environmental condition do not vary too much.

The effect of environmental conditions on measurement results should be evaluated globally for the whole sound pressure level measuring system in third-octave bands, based on manufactures' information concerning individual components of the system and determined in applicable normative documents, with actual range of variability of the static pressure, temperature, and humidity taken into account.

Measuring errors are also due to sensitivity to electromagnetic fields varying with the power grid frequency and to radio-frequency (RF) electromagnetic fields.

The effect of electromagnetic field varying with the AC power grid frequency or a RF interference should be taken into account only when tests conducted in the measuring environment confirm their presence and moreover, information concerning distribution and range of values characterizing intensities of the fields is available. Assessment of the effect of RF fields depends on the knowledge of sensitivity of the used measuring apparatus (and its individual components).

It is a well-known fact that the electromagnetic field varying with the power grid frequency may have an effect on the microphone which is usually determined as the sound pressure level equivalent to the signal generated at microphone terminals due to the effect of the field with standardized intensity in the direction of maximum response. For 1/4″ microphones, electromagnetic fields with rms intensity of 80 A/m, and frequency of 50 or 60 Hz, manufacturers declare typically the level from several dB to 30 dB (i.e., less than or comparable to the inherent noise level of typical microphone).

As far as the effect of radio-frequency electromagnetic fields is concerned, it is highly desirable to get to documented reliable tests (issued, e.g., as part of the type approval) evidencing conformity of the measuring apparatus with applicable requirements in the scope of electromagnetic compatibility.

From the point of view of measurement errors, an important issue is also sensitivity to mechanical vibrations. Sensitivity of microphones to mechanical vibrations emerges especially when they are exposed to vibrations in the direction perpendicular to the microphone membrane plane. Some manufacturers provide data in the form of the sound pressure level equivalent to the signal generated at microphone terminals as a result of mechanical vibrations of given acceleration value and the direction for which the maximum effect occurs. For 1/4-inch microphones, some manufacturers declare that the level is about 60 dB at exposure to axial vibrations characterized with the acceleration value of 1 m/s^2. Estimation of the uncertainty due to the effect of mechanical vibration must be preceded with a thorough investigation of their potential presence in the measuring environment.

An important factor is also sensitivity of measuring hardware to power supply voltage changeability. The related changes are more likely and can be more significant in case of prolonged measurements. Documented, reliable tests (carried out e.g. as part of the type approval procedure) provide data evidencing resistance of measuring apparatus to variation of power supply voltage.

8.5.6 THE RESEARCH METHOD

Any measuring method remains in close relationship with mode of the examined phenomenon. In case of the effect of exposure to ultrasonic noise, the model comprises information acquired by way of a survey and concerning the workstation to be tested, as well as preliminary information concerning typical process operations carried out at the workstation. Next, situations are identified which can be used to characterize worker's exposure to noise in the course of their working shift or any other reference period of time. Each of the identified situations, in the course of its occurrence is strictly connected with the worker's actual location in the course of their work, condition of noise sources having significant effect on the measured values, the situation duration period, and the sound pressure level averaged over the situation duration time. The set of all significant situations constitutes a model of the phenomenon of exposure.

Workstations involving exposure to ultrasonic noise are typically stationary posts, and operations carried out at such stands can usually be divided into clearly distinguishable time intervals. In such case, the most effective measuring method involves the procedure of division into tasks [ISO 9612:2009] which consists in analysis of the work and dividing it into a number of representative activities called *tasks* for which separate measurements are carried out.

In cases when the sound pressure level averaged for given situation (task) is determined based on several elementary measurements (samples), the actual value is unknown and its estimate is determined according to the formula

$$L_{p,\mathrm{eq},A,T_m} = 10\,\log10\left(\frac{1}{N}\sum_{n=1}^{N}10^{0.1\cdot L_{p,\mathrm{eq},A,T_{mn}}}\right)\mathrm{dB} \qquad (8.1)$$

where:

L_{p,eq,A,T_m} (dB): A-weighted time-averaged equivalent sound pressure level in the course of task m with duration T_m

n: number of samples for the task m

N: total number of samples for the task m

Equation 8.1 is a non-linear formula with respect of individual values used to calculate the final measurement result. In such case, ISO/IEC Guide 98-3:2008 recommends that when the uncertainty is determined by means of approximation with the use of Taylor series, higher-order terms should be taken into account. Determination of high-order terms is difficult both computationally and methodologically as far as their contribution to the uncertainty budget in the final estimate is concerned. For that reason, with intent to facilitate the process of uncertainty estimation in practice, it has been assumed that two values will contribute to the uncertainty of sampling—the standard deviation and the number of samples (a counterpart of the first-order term of Taylor series). It can be assumed that such approach, to an accuracy sufficient in practice, determines the uncertainty due to evaluation of the estimate of the equivalent sound pressure level for a given task from several elementary measurements.

8.5.7 Establishing the Minimum Number of Samples Required to Determine the Equivalent Sound Pressure Level

An issue important from the point of view of the measurement uncertainty consists in choosing the minimum number of samples necessary to determine the equivalent sound pressure level for given task carried out at a workstation. The effect of the number of samples and the standard deviation of their values on accuracy of evaluation of the equivalent sound pressure level estimate is shown in Figure 8.2.

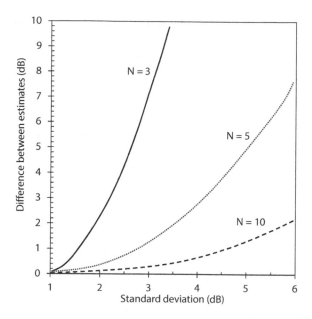

FIGURE 8.2   The effect of the number of samples N and the standard deviation on evaluation of the equivalent sound pressure level estimate. (From Thiery, L. and Ognedal, T., *Acta Acust. United Acust.*, 94, 331–334, 2008.)

It can therefore be concluded that there are no significant differences between two estimates determined based on a large number of samples with a small standard deviation value. However, the issue remains what minimum number of samples should be adopted in a measuring method which would enable one to determine the equivalent sound pressure level for a given task with adequate precision. Another issue important for practical reasons consists in determining what should be the maximum acceptable dispersion of values obtained from the measured samples. In the case of determining the equivalent sound pressure level from three samples, the difference between two estimates can be considered acceptable when the standard deviation for sample values does not exceed a maximum of about 1.5 dB. Therefore it is assumed that the minimum number of samples is 3, and the largest dispersion between sample values should not exceed 3 dB (the standard deviation value does not exceed about 1.5 dB in this case). The requirement defined in that way is consistent with requirements concerning noise measuring methods in the audible range.

An important element of any measuring method consists in adopting a minimum sample duration time. In audible noise measurements it is assumed that if the task duration is less than 5 min, then the duration of each elementary measurement should equal the duration of the task. In the case of tasks lasting longer, duration of each elementary measurement should be at least 5 min. The duration of each measurement may however be reduced if it becomes evident that the sound level is constant or repeatable or it can be concluded that the noise connected with given task has an insignificant effect on the total exposure to noise.

In our example, for the purpose of comparison and verification of requirements concerning the minimum duration of the elementary measurement, tests were carried out at workstations where the presence of ultrasonic noise was found [Radosz 2012], namely at DN5-75 drawing frame and Intersonic IS-4 ultrasonic washer operators' stands.

The testing method consisted in measuring the ultrasonic noise at the machine operator's workstations during three consecutive working shifts. In the course of measurements, the sound pressure level waveforms in third-octave bands were recorded in the range 10–40 kHz with the sampling frequency of 1 s.

Results of the measurements were subject to a simulation in the MATLAB® computing environment involving performance of 1000 iteration steps of the equivalent sound pressure level estimate determination procedure. The estimate value was determined from three or five samples taken randomly during three working shifts. The duration of each sample was either 15 s or 5 min.

As a result of the simulations, the 95% confidence interval was determined based on percentiles 2.5 and 97.5 from the obtained sound pressure level estimates. The analysis covered third-octave frequency bands in the aspect of admissible values specified in Table 8.2 and results are presented in Figures 8.3 and 8.4.

Results of the performed simulations proved that the ultrasonic noise measurement period as short as 15 s was insufficient to determine the equivalent sound pressure level for given tasks. The more appropriate measurement time seems to be the period of 5 min adopted in the noise measurement method (ISO 9612:2009), although in case of the ultrasonic washer and frequencies 12.5 and 20 kHz, the 95% confidence

---

TABLE 8.2

Accuracy of the Sampling Technique-Based Measuring Method Expressed by Means of 95% Confidence Interval for the Averaged Sound Pressure Level Estimate

			In Third-Octave Band with Mid-Band Frequency of (kHz)						
			10	12.5	16	20	25	31.5	40
Workstation	Measurement Duration Time (s)	Number of Samples	95% Confidence Interval for the Averaged Sound Pressure Level Estimate (dB)						
DN5-75 drawing	15	3	1.9	1.8	1.9	1.9	—	—	—
frame operator	15	6	0.8	0.9	0.9	0.9	—	—	—
	300	3	0.7	0.8	0.7	0.8	—	—	—
	300	6	0.6	0.6	0.6	0.7	—	—	—
Intersonic	15	3	1.4	2.3	1.3	2.9	—	1.2	1.3
IS-4 ultrasonic	15	6	0.9	1.5	1.0	2.0	—	0.9	0.9
washer	300	3	1.3	2.1	1.3	2.7	—	1.2	1.3
operator	300	6	0.8	1.1	0.9	2.0	—	0.8	0.8

Source: Radosz, J., *Noise Control Eng. J.*, 60, 645–654, 2012.

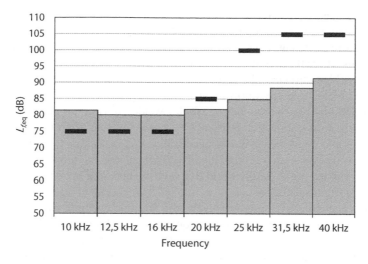

FIGURE 8.3   Results of ultrasonic noise measurements at the DN5-75 drawing frame operator's workstation (black bars mark admissible values).

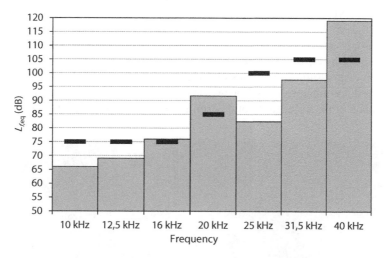

FIGURE 8.4   Results of ultrasonic noise measurements at the Intersonic IS-4 ultrasonic washer operator's workstation (black bars mark admissible values).

intervals exceeded 2 dB. Simulation results confirmed the relationship shown in Figure 8.2—the uncertainty due to the adopted sampling technique decreases with increasing number of elementary measurements.

8.5.8   IMPULSE NOISE IN THE RANGE 10–40 kHz

Another group of devices distinct from those discussed above in view of the waveform of the emitted ultrasonic noise signal are ultrasonic welding machines.

The devices generate the impulse noise (a specific sound comprising one or more acoustic events with duration shorter than 1 s each) of very high level exceeding admissible values [Smagowska 2013b]. The level of exposure to ultrasonic noise at workstations where the welders are installed is directly connected with the number of articles welded (directly related to the number of ultrasound pulses generated) in the course of staying at the workstation and sound pressure levels occurring in the course of welding. As the number of articles welded during the period of remaining within the workstation area is usually easy to determine, from the point of view of the ultrasonic noise measurement uncertainty it is necessary to gain knowledge about the waveforms of welding-related sound pulses. Examination of welding pulses was carried out at the Bielefeld 017 welding machine operator's workstation (Figure 8.5) [Radosz 2012].

In view of high sound pressure levels, the analysis encompassed all the third-octave bands from the ultrasonic noise frequency range (Figure 8.6). The waveforms were recorded in the course of a working shift at the sampling interval of 0.007 s.

Analysis of sound waveforms recorded for individual welding pulses revealed a large dispersion among the measured sound pressure levels. Confidence intervals based on percentiles 2.5 and 97.5 in extreme cases assumed the values of 9.1 dB for the frequency 25 kHz and 22.3 dB for the frequency 10 kHz (cf. Table 8.3). For the frequency 20 kHz, where the highest overruns of admissible values were observed, the value of 95% of the confidence interval was 11.4 dB. Results of the test prove that ultrasonic noise measurements at ultrasonic welding stations must be carried out with the utmost care when the related uncertainty is determined. The significant dispersion of pressure level values observed in the course of the welding process affects the determined values of the uncertainty of equivalent values and the process of evaluation of the uncertainty of determination of maximum level values.

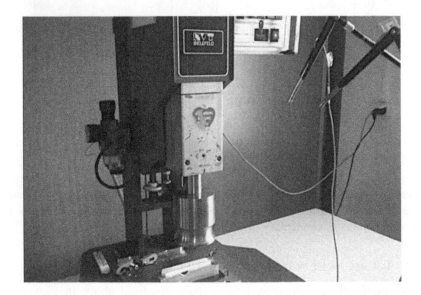

FIGURE 8.5 The examined Bielefeld 017 welding machine operator's workstation.

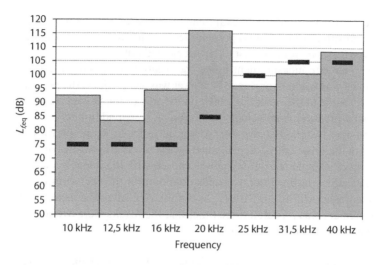

FIGURE 8.6 Results of ultrasonic noise measurements at the Bielefeld 017 welding machine operator's workstation (black bars mark admissible values).

TABLE 8.3

Extreme Values of the 95% Confidence Interval for Sound Pressure Levels Recorded During Pulses Generated by Bielefeld 017 Ultrasonic Welder

		In Third-Octave Band with Mid-Band Frequency of (kHz)						
		10	**12.5**	**16**	**20**	**25**	**31.5**	**40**
Workstation	**Percentile**	**Sound Pressure Level in the Course of a Welding Pulse (dB)**						
Bielefeld 017 welder	2.5	92.6	90.5	102.9	124.4	106.5	105.6	115.2
operator's station	97.5	114.9	105.0	114.0	135.8	115.6	123.1	132.0

Source: Radosz, J., *Noise Control Eng. J.*, 60, 645–654, 2012.

8.5.9 MICROPHONE LOCATION IN THE COURSE OF ULTRASONIC NOISE MEASUREMENTS

As was already noted in Sections 8.5.1 and 8.5.3, it is very important to ensure that in the course of measurements in the range of ultrasonic noise frequencies, the microphone has no protection grid installed and its axis is aimed at the noise emission source. Even when these requirements are met and the 10-cm distance from worker's ear is maintained or the measurements are taken at the point where the worker's head is usually situated [ISO 9612:2009], the measurement point may fall in one of many random places. That may be connected with such factors as the height of the worker who happened to be present at the workstation just at the moment of taking the measurement, the worker's moves within the workstation area, or imprecise location of the measurement point.

Under laboratory conditions, examination of the effect of microphone position on measured values of the sound pressure level generated at the Polsonic ultrasonic washer type 0.5 operator's workstation was carried out.

As the reference measurement point, the position at which the ultrasonic noise measurement was taken in presence of worker was selected, i.e. the point 10 cm away from the entrance to the outer ear canal, on the side of the worker's head exposed to higher sound pressure level values (see Figure 8.7). An additional reference microphone was also installed at the workstation in order to keep control of variations occurring in the noise emission generated by the ultrasonic washer over time.

Duration of a single measurement was chosen with the intent to ensure stabilization of indications in third-octave bands dominant in the spectrum of noise generated by the washer within ± 0.5 dB ($T_{e,i} = 5$ min). At each of the measurement points, the measurement was repeated three times and the results were averaged.

The subject of the test included:

1. The effect of the microphone orientation direction at the reference measurement point on the measured sound pressure level (Figure 8.8)—as the maximum, the microphone deviation of $\pm 20°$ relative to the reference direction (denoted as the direction 0) in two mutually perpendicular planes was assumed.
2. The effect of microphone position relative to the outer ear canal entrance (Figure 8.9) on the measured sound pressure level—as the maximum displacement of the microphone position, a move by ± 5 cm was adopted relative to the reference measurement point, in two mutually perpendicular planes.
3. The effect of microphone position in the working space area [Gedliczka 2001], (see, for example, Figure 8.10).

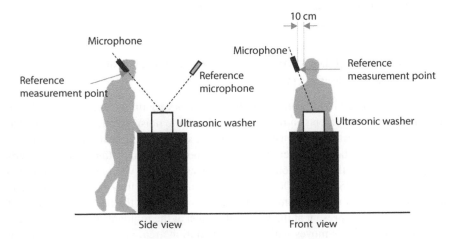

FIGURE 8.7 A schematic view of the ultrasonic washer operator's workstation.

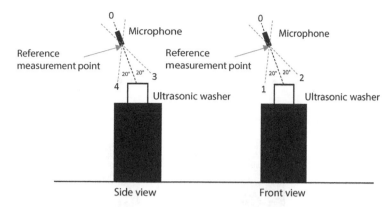

FIGURE 8.8 Microphone orientation directions in the course of the study on the effect of microphone orientation w reference measurement point; 0–4—microphone orientation directions at the reference measurement point.

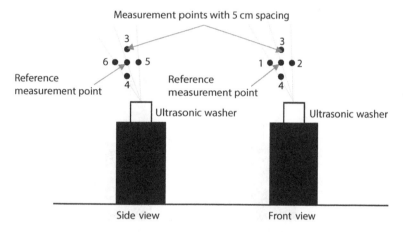

FIGURE 8.9 Situation of measurement points in the course of a study on the effect of microphone positioning relative to the outer ear canal entrance; 0–6—measurement taking points.

4. The effect of microphone positioning at the point where the center of worker's head is usually situated relative to the reference measurement point (Figure 8.11).

Results of measurements are presented in Tables 8.4 through 8.7. The measurements revealed that a microphone orientation direction change within the range of ±20° relative to the reference position has the least effect on the measured sound pressure level from among the considered measurement situations. In the dominant third-octave frequency band of the tested ultrasonic washer (40 kHz), the maximum difference between the measured values was 0.2 dB. In other frequency bands, the differences were found to be with in the limits 0.2–2.7 dB.

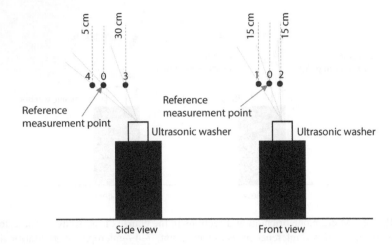

FIGURE 8.10 Situation of measurement points in the course of a study on the effect of microphone positioning accuracy in the working space area; 0–4—measurement taking points.

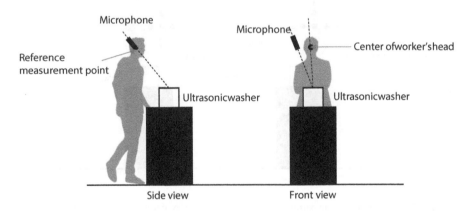

FIGURE 8.11 Situation of measurement points in the course of a study on the effect of microphone positioning at the point corresponding to worker's head center relative to the reference measurement point.

A more significant effect on the measured sound pressure level is connected with microphone location relative to the outer ear canal entrance (±5 cm). In the dominant third-octave frequency band of the examined ultrasonic washer (40 kHz), the maximum difference between the measured values respective to the reference measurement point was 2.8 dB. In the remaining frequency bands, the differences stayed within the range of 1.2–2.7 dB.

Significant differences in sound pressure levels occurred in the case of measurements at points of maximum torso inclination within the working space area. In the dominant third-octave band (40 kHz), the maximum difference between the

TABLE 8.4

Results of Measurements at the Reference Measurement Point for Different Microphone Orientation Directions

	In Third-Octave Band with Mid-Band Frequency of (kHz)						
	10	12.5	16	20	25	31.5	40
Microphone Position	Equivalent Sound Pressure Level in the Exposure Period (dB)						
0 (reference)	62.1	72.4	75.8	87.0	87.4	105.2	126.4
1	63.9	74.4	76.1	85.8	87.6	105.1	126.3
2	64.8	74.7	76.5	85.8	87.8	105.1	126.3
3	63.9	74.4	76.0	85.7	87.6	105.1	126.3
4	63.9	74.0	76.2	85.7	88.0	105.4	126.6
Maximum difference relative to the reference position (dB)	2.7	2.3	0.7	1.3	0.2	0.2	0.2
Standard deviation (dB)	1.0	0.9	0.3	0.5	0.2	0.1	0.1

Source: Radosz, J., *Noise Control Eng. J.*, 60, 645–654, 2012.

TABLE 8.5

Results of Measurements in the Vicinity of the Outer Ear Canal Entrance

	In Third-Octave Band with Mid-Band Frequency of (kHz)						
	10	12.5	16	20	25	31.5	40
Microphone Position	Equivalent Sound Pressure Level Over Exposure Period T_e (dB)						
0 (reference)	62.1	72.4	75.8	87.0	87.4	105.2	126.4
1	62.9	72.1	76.0	87.1	86.8	104.4	125.7
2	64.1	74.2	76.5	86.0	89.8	107.9	129.2
3	62.6	72.2	75.6	86.8	87.3	105.1	126.3
4	62.7	73.0	76.6	87.2	88.5	106.4	127.6
5	62.6	72.9	76.6	87.2	88.4	106.2	127.5
6	63.8	71.4	74.1	85.8	86.4	104.0	125.3
The maximum difference relative to the reference measurement point (dB)	2.0	1.8	1.7	1.2	2.4	2.7	2.8
Standard deviation σ (dB)	0.7	0.9	0.9	0.6	1.2	1.4	1.3

Source: Radosz, J., *Noise Control Eng. J.*, 60, 645–654, 2012.

measured values was 6.4 dB. In the remaining frequency bands, differences were found to fall into the interval 3.2–5.9 dB.

Large differences also occurred between results of measurements taken at the reference measurement point and the measurements made at the point where the worker's head is usually situated (head center). In the dominant third-octave frequency band of the tested ultrasonic washer (40 kHz), the maximum difference between the

TABLE 8.6

Results of Measurements at Points of Maximum Deflections of Worker's Torso Acceptable in the Course of Work

Microphone Position	In Third-Octave Band with Mid-Band Frequency of (kHz)						
	10	12.5	16	20	25	31.5	40
	Equivalent Sound Pressure Level Over Exposure Period T_e (dB)						
0 (reference)	62.1	72.4	75.8	87.0	87.4	105.2	126.4
1	64.1	72.0	75.9	84.8	87.1	104.4	125.6
2	64.9	72.9	77.3	88.3	90.9	109.4	130.7
3	67.2	76.3	80.5	90.2	93.1	111.1	132.8
4	63.8	71,4	74.1	85.8	86.4	104.0	125.3
The maximum difference relative to the reference measurement point (dB)	5.1	3.9	4.7	3.2	5.7	5.9	6.4
Standard deviation σ (dB)	1.9	1.9	2.4	2.1	2.9	3.2	3.4

Source: Radosz, J., *Noise Control Eng. J.*, 60, 645–654, 2012.

TABLE 8.7

Results of Measurements at the Point Corresponding to the Center of Worker's Head

Microphone Position	In Third-Octave Band with Mid-Band Frequency of (kHz)						
	10	12.5	16	20	25	31.5	40
	Equivalent Sound Pressure Level Over Exposure Period T_e (dB)						
0 (reference)	62.1	72.4	75.8	87.0	87.4	105.2	126.4
Worker's head center	64.9	73.0	77.4	88.4	90.9	109.4	130.8
The maximum relative to the reference measurement point (dB)	2.7	0.6	1.6	1.4	3.5	4.3	4.4
Standard deviation (dB)	1.9	0.4	1.1	1.0	2.5	3.0	3.1

Source: Radosz, J., *Noise Control Eng. J.*, 60, 645–654, 2012.

measured values was 4.4 dB. In other frequency bands, the differences were found to be with in the limits of 0.6–4.3 dB.

8.6   THE MEASUREMENT METHODOLOGY

The published ISO 9612:1997 standard included methodology for ultrasonic noise measurements, but it was replaced with its currently valid issue (i.e., ISO 9612:2009). The scope of the standard in its present form does not include ultrasonic noise measurements. In the light of the ISO 9612:1997 standard, ultrasonic noise measurements should be carried out with the use of third-octave filters or narrow-band filters with mid-band frequencies corresponding to the range of frequencies found in the examined noise spectrum. For the measurements, it was recommended to use a sound level meter or an integrating sound level meter of accuracy class 1 with known frequency characteristic above 12 kHz. The measurement time intervals were recommended to be selected in a way ensuring that the obtained results were representative for a long averaging period. The standard ISO 9612:1997 defines also a method for determination of the measurement uncertainty; however, it was published mainly for informative purposes and was inconsistent with requirements set out in ISO/IEC Guide 98-3:2008. Moreover, the method did not consider the specificity of the ultrasonic noise frequency range to a sufficient degree.

Ultrasonic noise measurements are carried out at points where workers are located while performing their tasks and include determination of individual activities (tasks) carried out by them and representative conditions of operation of devices, machines, and tools which are sources of the noise. Workstations involving exposure to ultrasonic noise, especially ultrasonic machine tools used in the production process (ultrasonic welding machines, ultrasonic washers, ultrasonic erosion machines) are usually stationary work stands and tasks realized at them can be divided into clearly distinguishable time intervals depending on the mode of operation of the devices. In such case, the most effective measuring method is that which includes division of the work carried out into tasks proposed in the standard ISO 9612:2009 adapted to the examined frequency range. The method consists in analyzing the work and dividing it into a number of representative activities (tasks) for which separate measurements are carried out.

A matter of special importance when taking measurements in the range of ultrasonic noise frequencies is the position of the microphone. Even in case of maintaining the 10-cm distance from the worker's ear or taking a measurement at the point where the worker's head can be found normally [ISO 9612:2009], the measurement point can fall in many random places. That may be connected with, for example, the height of the worker who happened to be present at the workstation just at the moment of taking the measurement, the worker's moves within the workstation area, or an insufficiently precise location of the measurement point. In the range 10–40 kHz, the maximum difference in the measured values, at different points of the workstation, relative to the reference measurement point can exceed 6 dB [Radosz 2012]. Microphone position in the vicinity of the outer ear canal entrance also has an important effect in the measurement results—in case of displacement by ±5 cm in any direction, the maximum difference between results of the measurement relative to the reference point can be as high as 3 dB [Radosz 2012].

In view of the above, it is recommended that in the course of taking the measurements, the directivity pattern of the ultrasonic noise source be determined or, at the minimum, examination must be undertaken of the sound field around the source to at least such degree of precision which ensures that the microphone is pointed in direction corresponding to maximum noise emitted by the source.

The period of work carried out at the examined workstation should be divided into time intervals representing duration of individual tasks distinguishable from the point of view of exposure to ultrasonic noise. All the significant ultrasonic noise contributions should be taken into account and the combined duration time of individual tasks should cover the entire working shift.

In many cases, the exposure time is directly connected with the number of cycles of the actually performed work. A good example is ultrasonic welder operators' workstations where exposure to ultrasonic noise is closely related to the number of items welded in the course of the operator's shift at the workstation and with the sound pressure levels in the course of welding.

In such cases, the most accurate method for estimating the duration of a given exposure interval (task) consists in referring to production schedules. If such an approach is adopted, then it is necessary to take several elementary measurements under the assumption that a specific number of welding cycles is registered. From duration of given number of cycles, the average value of single cycle duration times is calculated to determine the welding cycle duration $T_{m,j}$. Next, the total duration of welding cycles effected in the course of working shift T_m is determined based on number of cycles set out in the production schedule. Given the corresponding the values, it is possible to determine the total task duration time as

$$T_m = \frac{N}{J} \sum_{j=1}^{J} \frac{T_{mj}}{n_j} \text{ h, min or s} \tag{8.2}$$

where:

T_{mj} (h, min, or s): duration of j-th observation (sample) for m-th task
n_j: number of welding cycles in the course of j-th observation
J: total number of measurements taken for m-th task
N: total number of articles fabricated for m-th task as determined by the production schedule

In case of other measurement-related issues (e.g., the measurement duration time, number of measurements, etc.), it is possible to adapt accordingly the relevant recommendations of ISO 9612:2009 standard.

An alternative method for determination of occupational exposure to ultrasonic noise is the approach in which measurements of individual acoustic events are carried out. The method is based on the fact that selected tasks or acoustic situations (e.g., operating an ultrasonic welding machine) are characterized not by means of time intervals for which they last but rather in terms of type and number of acoustic events occurring in that period (e.g., the number of items welded per working shift). The measurement start and end moments should be chosen, with specificity of the workstation taken into account, in a way that ensures that the whole of a characteristic repeatable cycle of variability of ultrasonic noise is covered, closely connected with the task m carried out by the worker. The quantity suitable to characterize individual acoustic events is the sound exposure level (denoted SEL or L_{AE}) measured with a meter/analyzer simultaneously with the equivalent level in i-th frequency band $L_{p,eq,i}$. Definition of the sound exposure level is analogous to that of the equivalent level. The only difference consists in that calculation of the value in question, instead of using the observation time T, is based on the conventional reference time of 1 s:

$$SEL_i = 10\log_{10}\left(\{t\}10^{0.1 \cdot L_{p,eq,i,T}}\right) = L_{p,eq,i,T} + 10\log_{10}\{t\} \text{ dB} \qquad (8.3)$$

where:

t (s): the measurement duration time

$L_{p,eq,i,T}$ (dB): the equivalent sound pressure level in i-th frequency band for the observation time T

One can therefore assume that

$$SEL_{i,m} = L_{p,eq,i,T_m} + 10 \log_{10}\langle t_m \rangle \text{ dB} \qquad (8.4)$$

where:

$SEL_{i,m}$ (dB): is the level of exposure to noise referred to 1 s in i-th frequency band measured in the course of one of N_m events making up the task m

L_{i,eq,T_m} (dB): equivalent sound pressure level in i-th frequency band for the time T_m of observation of task m

$\langle t_m \rangle$ (s): average time of measurement carried out in the course of task m

By substituting the reference time of 8 h (28,800 s), the following formula for the contribution of noise attributable to task m at the daily level of exposure to noise is obtained:

$$L_{p,i,eq,8 \text{ h},m} = SEL_m + 10 \log_{10}\langle N_m \rangle dB - 10 \log_{10}(28,800) \text{ dB} \qquad (8.5)$$

where:

SEL_m (dB): exposure to noise level standardized to 1 s measured in the course of one of N_m events making up the task m

$\langle N_m \rangle$: the average number of acoustic events making up the task m

The presented measuring method applicable to individual acoustic events may be considered an elaboration of the Strategy 1 presented in ISO 9612. The sound exposure levels SEL are measured for individual acoustic events. The SEL measuring time for each isolated acoustic event must not be less that the acoustical duration of the event. SEL values should be measured with the use of methods and at moments of the working shift selected in a way ensuring that the effect of acoustic events or tasks identified in the worker's exposure model other than those examined is eliminated.

8.7 CORRECTION OF MEASUREMENT RESULTS

In view of the significant effect of using the microphone protection grid and metrological characteristics of the apparatus on measurement results, the acoustic (equivalent and maximum) pressure level in i-th frequency band should be subjected to correction according to the formula:

$$L_{eq,i} = L'_{eq,i} + K_{app,i} + K_{grid,i} \text{ dB}$$ (8.6)

where:

$L'_{eq,i}$ (dB): meter/analyzer indication in i-th third-octave band

$K_{app,i}$ (dB): correction for combined effect of metrological characteristics of the apparatus on the measurement result; the correction is determined in the course of metrological inspection for each third-octave band as the sum of the frequency characteristic values of microphone (taken with opposite sign) and the relative attenuation of the filter, the figures being taken from calibration certificates

$K_{grid,i}$ (dB): correction for the effect of using the microphone protection grid on the measurement result

Where it is impracticable to use a microphone without a protection grid, it is necessary to apply the correction provided by the microphone manufacturer. If the manufacture's data on the effect of a protection grid are unavailable, the relevant correction should be determined experimentally.

8.8 THE MEASUREMENT UNCERTAINTY ESTIMATION

According to ISO/IEC Guide 98-3, when a result uncertainty is estimated, all component uncertainties significant in the given situation should be taken into account with the use of appropriate methods of analysis. Any decision whether or not a specific uncertainty component can be omitted must take into account: relative amplitudes of the largest and the smallest component; the effect of individual components on the estimated uncertainty; substantiation of the degree of precision adopted in determination of the uncertainty; customer requirements; applicable legal instruments; and other third-party criteria.

It is also possible to assume an alternative approach to calculation of the uncertainty and determine it, for instance, based on relative acoustic pressure values expressed in pascals. That way, the problem of performing arithmetic calculations on quantities expressed in decibels is circumvented. Where methods of that type are used, they should meet the requirements set out in the standard ISO/IEC Guide 98-3. It is also necessary to prove that the adopted method does not lead to underestimation of uncertainties. The method should be indicated in the measurement report.

Bearing in mind that the input values to the uncertainty budget are not intercorrelated, the combined standard uncertainty of determination of the equivalent sound pressure level in i-th third-octave band $u_i(L_{p,eq,8\,h,i})$, according to ISO/IEC

Guide 98-3, is calculated based on numerical values of individual component uncertainties from the formula:

$$u_i\left(L_{p,\text{eq},8\,\text{h},i}\right) = \sqrt{\sum_{m=1}^{M} c_{i,m}^2 \left(u_{\text{disp},i,m}^2 + u_{\text{app},i}^2 + u_{\text{mic},i}^2\right)} \ \text{dB} \tag{8.7}$$

where:

$c_{i,m}$: sensitivity coefficient in i-th third-octave band for m-th task

$u_{\text{disp},i,m}$ (dB): standard uncertainty due to dispersion of sample values in i-th third-octave band for m-th task

$u_{\text{app},i}$ (dB): the standard uncertainty due to the measuring apparatus in i-th third-octave band (see Table 8.1)

$u_{\text{mic},i}$ (dB): the standard uncertainty due to microphone position in i-th third-octave band (see Table 8.2)

m: task number

M: total number of tasks

The sensitivity coefficients determine the degree to which a given exposure interval affects the determined value of the equivalent sound pressure level referred to an 8-hour or weekly working time frame. The sensitivity coefficient $c_{i,m}$ is calculated from the formula:

$$c_{i,m} = \left(T_m / T_0\right) 10^{0.1(L_{p,\text{eq},T_m,i} - L_{p,\text{eq},8\text{h},i})} \tag{8.8}$$

where:

$L_{p,\text{eq},T_m,i}$ (dB): equivalent sound pressure level in i-th frequency band for m-th task

$L_{p,\text{eq},8\,\text{h},i}$ (dB): equivalent sound pressure level in i-th third-octave band standardized to 8-hour daily working time

T_m (s): task duration time

T_0 (h): reference time with $T_0 = 8$ h

In the adopted measuring method, the standard uncertainty due to dispersion of sample values $u_{i,m}$ for given exposure interval (task) is determined from the formula:

$$u_{\text{disp},i,m} = \sqrt{\frac{1}{J(J-1)} \sum_{j=1}^{J} \left(L_{p,\text{eq},T_{mj},i} - \left\langle L_{p,\text{eq},T_m,i}\right\rangle\right)^2} \ \text{dB} \tag{8.9}$$

where:

$L_{p,\text{eq},T_{mj},i}$ (dB): equivalent sound pressure level in i-th third-octave band for j-th measurement (sample) and m-th task

J: number of measurements taken for m-th task

$\langle L_{i,p,\text{eq},T_m}\rangle$ (dB): arithmetic average of equivalent sound pressure levels in i-th frequency band for m-th task

Values of the standard uncertainty due to the measuring apparatus $u_{\text{app},i}$ as determined empirically for the range 10–40 kHz are listed in Table 8.8 [Radosz 2014].

TABLE 8.8
The Standard Uncertainty Due to the Measuring Apparatus

Third-octave band with mid-band frequency (kHz)	10	12.5	16	20	25	31.5	40
Standard uncertainty $u_{\text{app},i}$ (dB)	0.45	0.46	0.46	0.47	0.47	0.49	0.50

TABLE 8.9
The Standard Uncertainty Due to Microphone Position

Third-octave band with mid-band frequency of (kHz)	10	12.5	16	20	25	31.5	40
Standard uncertainty $u_{\text{mic},i}$ (dB)	0.7	0.9	0.9	1.0	1.2	1.3	1.5

Values of the standard uncertainty due to microphone position for the frequency range 10–40 kHz are given in Table 8.9 [Radosz 2012]. If the microphone is placed too close to the worker's body, the measurement uncertainty is due to screening and reflection of sound. In case of measurements taken without the presence of operators at their workstation, the uncertainty is due to microphone position(s) which imprecisely correspond(s) to actual positions of the worker's head.

Finally, the expanded uncertainty of the equivalent sound pressure level in i-th third-octave band is determined from the formula:

$$U_i\left(L_{p,\text{eq,8 h},i}\right) = 1.65 u_i\left(L_{p,\text{eq,8 h},i}\right) \text{ dB} \tag{8.10}$$

where:

$u_i(L_{p,\text{eq,8 h},i})$ (dB): combined standard uncertainty of equivalent sound pressure level in i-th third-octave band

As the determined ultrasonic noise level values are to be used for the purpose of comparing them with admissible values, a one-sided 95-percent confidence interval was assumed in the present procedure which corresponds to the coverage factor $k = 1.65$ (on the analogy of ISO 9612). That means that 95% of values are below the lower confidence interval limit. The expanded uncertainty is presented as a separate figure.

Where necessary, the uncertainty budget may be completed with the standard uncertainty u_{i,T_m} due to estimation of duration time T_m of m-th task. In such case, Equation 8.7 should be extended to the following form (cf. ISO 9612:2011):

$$u_i\left(L_{p,\text{eq,8 h},i}\right) = \sqrt{\sum_{m=1}^{M} c_{i,m}^2 (u_{\text{disp},i,m}^2 + u_{\text{app},i}^2 + u_{\text{mic},i}^2) + \left(4.34\frac{c_{i,m}}{T_m} u_{i,T_m}\right)^2} \text{ dB} \tag{8.11}$$

where:

$$u_{i,T_m} = \sqrt{\frac{1}{J(J-1)} \sum_{j=1}^{J} (T_{m,j} - T_m)^2} \ \text{dB} \qquad (8.12)$$

and J is the total number of observations of the m-th task duration time.

If the task duration time is the result of work schedule analysis, then an estimate of the above standard uncertainty can be assumed in the form:

$$u_{i,T_m} = 0.5(T_{max} - T_{min}) \ \text{dB} \qquad (8.13)$$

In some cases, it may prove to be necessary to determine the peak or maximum values of the sound pressure level in third-octave bands. To determine maximum uncertainties for sound pressure levels in i-th third-octave band, one can use the tolerance limits method [Rudno-Rudziński 2007a, 2007b]. In that method, the upper tolerance limit for the maximum (or peak) sound pressure levels is determined in i-th frequency band to a definite probability. That means that with an assumed likelihood, an area (uncertainty) is established in which values larger than definite maximum or peak values may occur. Like in the case of equivalent sound pressure levels, the 95-percent level of confidence is assumed. The upper tolerance limit $L_{p,max(0.95),i}$ is calculated from the formula:

$$L_{p,max(0.95),i} = \langle L_{p,max,i} \rangle + k_1 \cdot u_{max,i} \ \text{dB} \qquad (8.14)$$

where:

$\langle L_{p,max,i} \rangle$ (dB): arithmetic average of maximum (or peak) sound pressure levels in i-th frequency band

$u_{max,i}$ (dB): the standard uncertainty due to dispersion of sample values in i-th frequency band

k_1: coverage factor for noncentral Student's t-distribution

When the combined standard uncertainty for maximum or peak values of the sound pressure level in the uncertainty budget is determined, the sensitivity coefficients are not taken into account as the maximum level is defined as the highest of the obtained maximum values from among recorded elementary measurements (samples). The uncertainty of maximum or peak levels is determined for the tasks for which the highest of recorded values in given third-octave band occurred.

Values of the standard uncertainty due to the measuring apparatus in the range 10–40 kHz are listed in Table 8.10 [Radosz 2014].

TABLE 8.10

Values of the Standard Uncertainty Due to the Measuring Apparatus $u_{app,max,i}$ for Maximum or Peak Levels

Third-octave band mid-band frequency (kHz)	10	12.5	16	20	25	31.5	40
Standard uncertainty $u_{app,max,i}$	0.44	0.45	0.45	0.45	0.46	0.48	0.49

8.9 SUMMARY

The subject matter covered in this chapter pertained to ultrasonic noise measurements in the work environment. In view of unquestionably evidenced harmfulness of noise components emitted in frequency range above 20 kHz, there is a disturbing shortage of unambiguous and exhaustive information about factors affecting results of sound pressure level measurements. There are also no currently applicable and acceptable international standards concerning the methodology of ultrasonic noise measurements at workstations. Analysis of regulations concerning ultrasonic noise at workstations, measuring methods, metrological requirements specific for the measuring apparatus, and detailed identification of factors affecting measurement results reveals the necessity to develop a specific method for ultrasonic noise measurements. The ultrasonic noise measuring method proposed in this chapter includes, in particular: requirements concerning the measuring apparatus and scheduled metrological inspection; requirements applicable to the measuring environment (temperature, humidity, and static pressure); and description of conduct in the course of measuring sessions. In the method, the use of necessary corrections to measurement results and the procedure of determination of uncertainties are also taken into account with emphasis put on maintaining consistence with standards applicable in the range of audible acoustic and infrasonic frequencies.

REFERENCES

Acton, W. I. 1973. The effects of airborne ultrasound and near-ultrasound. *Proceedings of International Congress on Noise as a Public Health Problem*, Dubrovnik, Croatia, pp. 13–18 May, 1973.

Acton, W. I. 1974. The effects of industrial airborne ultrasound on humans. *Ultrasonics* 12:124–128. DOI:10.1016/0041-624X(74)90069-9.

Brüel & Kjær. *Microphone Handbook*, Vol. 1, *Theory*. Technical Documentation, BE 1447–11. 1996. Nærum: Brüel & Kjær.

Gedliczka, A. 2001. *Atlas Miar Człowieka*. [Atlas of human measures]. Warszawa: CIOP-PIB.

IEC 60942:2003. Electroacoustics—Sound calibrators. 2003. International Electrotechnical Commission. Geneva, Switzerland.

IEC 61260-1:2014. Electroacoustics—Octave-band and fractional-octave-band filters—Part 1: Specifications. 2014. International Electrotechnical Commission. Geneva, Switzerland.

IEC 61672-1:2013. Electroacoustics—Sound level meters—Part 1: Specifications. 2013. International Electrotechnical Commission. Geneva, Switzerland.

IEC 61672-3:2013. Electroacoustics—Sound level meters—Part 3: Periodic tests. 2013. International Electrotechnical Commission. Geneva, Switzerland.

ISO 9612:1997. Acoustics Guidelines for the measurement and assessment of exposure to noise in a working environment. 1997. International Organization for Standardization. Geneva, Switzerland.

ISO 9612:2009. Acoustics—Determination of occupational noise exposure—Engineering method. 2009. International Organization for Standardization. Geneva, Switzerland.

ISO/IEC Guide 98-3:2008. Uncertainty of measurement—Part 3: Guide to the expression of uncertainty in measurement (GUM: 1995). 2008. International Organization for Standardization. Geneva, Switzerland.

Koton, J. 1986. *Ultradźwięki* [Ultrasound]. Warszawa: Instytut Wydawniczy Związków Zawodowych.

Lawton, B. W. 2013. Exposure limits for airborne sound of very high frequency and ultrasonic frequency (ISVR Technical Report, 334). Southampton: University of Southampton.

Mikulski, W., and J. Radosz. 2010. The measurements of ultrasonic noise sources in the frequency range from 10 kHz to 40 kHz. *Proceedings of 17th International Congress on Sound and Vibration ICSV 2010*, Vol. 5, 3656–3663, Cairo, Egypt, 18 July–22 July.

Pawlaczyk-Łuszczyńska, M., A. Dudarewicz, and M. Śliwińska-Kowalska. 2007. Sources of occupational exposure to ultrasonic noise. *Med Pr* 58(2):105–116.

Radosz, J. 2012. Methodology issues of ultrasonic noise exposure assessment. *Noise Control Eng J* 60(6):645–654. DOI:10.3397/1.3701038.

Radosz, J. 2014. Uncertainty due to instrumentation for sound pressure level measurement in high frequency range. *Noise Control Eng J* 62:(4):186–195. DOI:10.3397/1/376219.

Radosz, J., and D. Pleban. 2018. Ultrasonic noise measurements in the work environment. *J Acoust Soc Am* 144(4):2532–2538. DOI:10.1121/1.5063812.

Rudno-Rudziński, K. 2007a. Ocena niepewności pomiaru poziomów maksymalnych hałasu w środowisku pracy [Evaluation of the maximum sound pressure level measurement uncertainty in the work environment]. *Proceedings of XXXV Zimowa Szkoła Zwalczania Zagrożeń Wibroakustycznych*, Gliwice-Ustroń, Poland, 26 February–2 March, 2007.

Rudno-Rudziński, K. 2007b. Why and how to determine peak workplace noise pressure measurement uncertainty. *Proceedings of 9th International Congress on Acoustics*. Madrid, Spain, 2–7 September, 2007.

Smagowska, B. 2013a. An objective and subjective study of noise exposure within the frequency range from 10 to 40 kHz. *Arch Acoust* 38(4):559–563. DOI:10.2478/aoa-2013-0066.

Smagowska, B. 2013b. Ultrasonic noise sources in the work environment. *Arch Acoust* 38(2):169–176. DOI:10.2478/aoa-2013-0019.

Smagowska, B., and M. Pawlaczyk-Łuszczyńska. 2013. Effects of action of ultrasonic noise on the human body – A bibliographic review. *Int J Occup Saf Ergon* 19(2):195–202. DOI:10.1080/10803548.2013.11076978.

Śliwiński, A. 2001. *Ultradźwięki i ich zastosowania* [Ultrasound and its application]. Warszawa: Wydawnictwa Naukowo-Techniczne.

Śliwiński, A. 2016. On the noise hazard assessment within the intermediate range of the high audible and the low ultrasonic frequencies. *Arch Acoust* 41(2):331–338. DOI:10.1515/aoa-2016-0034.

Thiery, L., and T. Ognedal. 2008. Note about the statistical background of the methods used in ISO/DIS 9612 to estimate the uncertainty of occupational noise exposure measurements. *Acta Acust United Acust* 94(2):331–334. DOI:10.3813/AAA.918037.

9 Studies on Sound Insulation of Enclosures in the 10–40 kHz Frequency Range

Witold Mikulski

CONTENTS

9.1 INTRODUCTION

The basic parameter used to characterize the effectiveness of acoustic protection provided by sound-insulating enclosures of machines and devices is the so-called sound power insulation $D_{W,f}$ [Engel and Sikora 1998; EN ISO 15667:2000; EN ISO 11546-2:2009; Dobrucki et al. 2010; Pleban and Mikulski 2015, 2018; Mikulski 2019] which for a given enclosure is defined as the reduction in sound power level achieved due to the enclosure (i.e., the difference between the sound power levels from a reference sound source without enclosure and the same source provided with the enclosure). This approach is also known as the insertion loss method [Pleban and Mikulski 2015, 2018; EN ISO 11546-2:2009]. Currently, the sound insulation of sound-insulating enclosures is determined for frequencies up to about 10 kHz according to EN ISO 11546-2:2009 [EN ISO 15667:2000; Pleban and Mikulski 2015, 2018], whereas the sound power level of sound sources is determined in the frequency band up to 10 kHz (as per EN ISO 3744:2010 and EN ISO 3746:2010) or since 2015, up to 20 kHz (according to EN ISO 9295:2015). This is probably the reason why European Union Directives 2006/42/EC, 2000/14/EC, and 2005/88/EC impose the obligation to determine noise emissions for machines and devices, but only in frequency ranges of up to 20 kHz. From this it follows that in case of machines and devices emitting sound energy mainly in the frequency range above 20 kHz (such as, ultrasonic washers, ultrasonic welders, ultrasonic erosion machines, etc. [Pawlaczyk-Łuszczyńska et al. 2007; Mikulski and Radosz 2010; Smagowska 2013b), the contribution of the hazard introduced by these equipment items to the work environment remains neglected. At the same time, it is a well-known fact reflected both in reports concerning studies on noise at workstations [Grzesik and Pluta 1983; Holmberg et al. 1995; Smagowska and Pawlaczyk-Łuszczyńska 2013; Kling et al. 2015; Śliwiński 2016; Pleban et al. 2018; Radosz and Pleban 2018] and in applicable regulations (including Polish ones) that the noise generated in the frequency range 10–40 kHz poses an important hazard to human health [Lawton 2013; Smagowska 2013a; Smagowska and Pawlaczyk-Łuszczyńska 2013; Śliwiński 2013; Radosz 2014; Pleban et al. 2018; Radosz and Pleban 2018. In Poland, the airborne sound in the frequency range 10–40 kHz is called the *ultrasonic noise*.

The 10–40 kHz frequency range is a commonly adopted abbreviated term as it comprises one-third octave frequency bands with mid-band frequencies 10 kHz, 12.5 kHz, 16 kHz, 20 kHz, 25 kHz, 31.5 kHz, and 40 kHz. In fact, the band covers the frequency range of 8.91–44.88 kHz. The analogous frequency span referred to in short as the 20–40 kHz range, covers actually the frequencies included between 17.82 kHz and 44.88 kHz.

Knowledge of the nature of sound source emission and determination of sound-insulating properties of sound-insulating enclosures in the frequency range of 10–40 kHz would enable one to design and apply effective sound-absorbing enclosures and thus reduce exposure of humans to that type of noise.

The sound power insulation of a sound-insulating enclosure depends on its structure, sound-insulating properties of individual structural components of the enclosure (including its walls), acoustic tightness of joints between component elements of the enclosure, possible presence of orifices (e.g., ventilation holes), and sound-absorbing properties of inner surface of the structure (including materials used for inner lining of the enclosure) [Mikulski 2013, 2019; Pleban 2013; Pleban and Mikulski 2018]. As the majority of sound-insulating enclosures used to neutralize ultrasonic noise sources are lightweight rigid-frame structures, that type of enclosures will be the subject of further detailed considerations.

This chapter is a review of numerical and combined computational-measuring methods for determination of the sound power level of a source and the sound insulation for sound-insulating enclosures in the frequency range of 20–40 kHz. The methods are based on measurements of the sound pressure level. The example of a plywood rigid-frame enclosure was used to examine the effect of thickness of its walls and of the sound-absorbing material used for the enclosure lining on the sound power insulation characterizing the enclosure.

9.2 A MEASURING-COMPUTATIONAL METHOD FOR DETERMINING THE SOUND POWER LEVEL OF SOURCE IN THE 10–40 kHz FREQUENCY RANGE

9.2.1 A GENERAL DESCRIPTION OF THE MEASURING-COMPUTATIONAL METHOD FOR DETERMINATION OF THE SOUND POWER LEVEL OF SOURCE IN THE 10–40 kHz FREQUENCY RANGE

The measuring-computational method for determination of the sound power level consists in taking measurements of the sound pressure level generated by the sound source in the 10–40 kHz frequency range (with the enclosure with source inside also considered as a sound source) and using the results to calculate the sound power level. In the following, the method will be referred to as the "measuring method." The method is based on standard sound power level of source measuring methods applicable to frequencies below 10 kHz (described in EN ISO 3744:2010 and EN ISO 3746:2010) and below 20 kHz (described in EN ISO 9295:2015). For frequencies higher than 20 kHz, it is necessary to take into account a number of factors for which lower frequencies could not be neglected or which play a much more important role at frequencies above 20 kHz [ISO 9613-1:1993; EN ISO 9295:2015; Pleban and Mikulski 2015]. The following may be numbered among such factors: high directivity characterizing sources of ultrasonic noise (requiring an increase in the number of measuring points provided for in the method) [Pawlaczyk-Łuszczyńska et al. 2007; Mikulski and Radosz 2010; EN ISO 9295:2015]; significant effect of sound attenuation in air on sound propagation [ISO 9613-2:1996; Mikulski and Radosz 2010; Mikulski, 2013; EN ISO 9295:2015] (necessity to compensate for the attenuation in measurement results by means of adopting, among other things, a hemispherical measurement surface [EN ISO 3744:2010; EN ISO 3746:2010; Mikulski and Radosz 2010; EN ISO 9295:2015] and an appropriate correction); and the necessity to ensure an adequate ratio of the signal of source (in enclosure) to the background noise.

The sound power level testing environment is a half-space limited by a sound-reflecting plane. A sound source placed on a sound-reflecting surface radiates the acoustic energy into the half-space (i.e., into the solid angle of 2π steradians) and the sound wave front surface area at the distance of r from the source is $2\pi r^2$. This does not necessarily mean that radiation of sound energy from the source is omnidirectional (in the considered frequency range of 10–40 kHz, omnidirectional radiation of sound energy is virtually impracticable). In the portion of space in which measurements are taken, the sound energy propagated directly from the source should dominate. In the measuring environment, the following contribute to the resultant sound pressure level on the measurement surface: the sound wave coming directly from the source; the sound wave emitted by the source and then reflected from and diffracted on obstacles; and the sound background. The effect of reflected waves is eliminated from the measurement results (that does not apply to the effect of the wave reflected from the rigid ground which is already taken into account in the correction for the measurement surface area) and the effect of background noise. The effect of reflected waves is taken into account by introducing an environmental correction, whereas the effect of background noise is represented by another specifically defined additive correction. The sound pressure level values measured on the measurement surface depend also on attenuation of sound in air. As the result of a sound power level measurement must not depend on the measurement surface (except for the correction for the correction proportional to $10\log\{S\}$, where S is the measurement surface area), the correction $K_{air,f}$ is introduced which eliminates the effect of sound wave attenuation in air. The correction depends on the distance of the measurement point from the source; therefore to simplify relevant calculations, all measurement points are situated at the same distance from source (i.e., they are distributed on surface of a hemisphere). Thus, the correction due to sound absorption in air has the same value for all measurement points.

9.2.2 The Sound Power Level of Source Measuring Procedure in the 10–40 kHz Frequency Range

Noise emission from sound sources in the 10–40 kHz frequency range is determined usually in one-third octave frequency bands with mid-band frequencies of 10 kHz, 12.5 kHz, 16 kHz, 20 kHz, 25 kHz, 31.5 kHz, and 40 kHz, so all calculations should be independently performed for all the criteria (including those concerning the testing environment and the background noise) applied to each of the frequency bands and therefore no functional dependence on frequency will be indicated explicitly hereinafter unless necessary in specific situations.

9.2.2.1 The Choice of the Measurement Surface

The measurement surface is a geometrical surface defined in space around a sound source limited from the bottom with a sound-reflecting ground surface. Inside the volume limited by the measurement surface and the ground plane, there is the sound source and the so-called reference cuboid—the smallest rectangular prism circumscribed on the sound source. Geometrically, the measurement surface is a hemisphere with the radius r called the measurement radius (Figure 9.1). Geometrically, condition $r \geq 2d_0$ must be met where d_0 is the distance from the base center point to the upper vertex of the reference cuboid (see Figure 9.5).

The radius r of the hemispherical measurement surface (the measurement radius) should be assumed as equaling 1 m. In the proposed procedure, values of some parameters (e.g., the number of measurement points and the radius and surface area of the measurement surface) are at first adopted provisionally. The next step consists in verifying, by means of calculations and/or measurements, whether the parameters are adopted correctly (i.e., with the adopted values of the parameters, all the

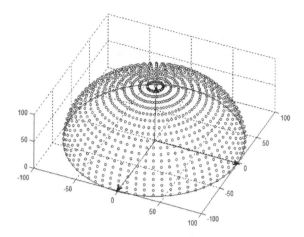

FIGURE 9.1 Hemispherical measurement surface in the proposed method with measurement points marked on it (the number of measurement points exceeding 1000). (From Mikulski, W. et al., Opracowanie wytycznych akustycznych projektowania obudów dźwiękoizolacyjnych przemysłowych technologicznych źródeł hałasu ultradźwiękowego. Opracowanie modelu obudowy dźwiękoizolacyjnej na wybrane technologiczne źródło hałasu ultradźwiękowego, Warszawa, CIOP-PIB, 2019.)

criteria concerning, for instance, the test environment are met). If not, the provisionally adopted values of the parameters (e.g., the measuring radius r) are modified until the assumed criteria are fulfilled.

In case of finding that the conditions for the testing environment are not met and the value of the environmental correction $K_{env,i}$ (denoted $K_{2,i}$ in the standard EN ISO 3744:2010) is too large, i.e., $K_{env,i} > 4$ dB in a specific i-th one-third octave frequency band, the measurement radius r should be decreased to less than 1 m, but still provided that $r \geq 2d_0$ and $r \geq 0.5$ m. The prerequisite for application of the method is that the measurement radius r must be selected so that $K_{env,i} \leq 4$ dB. Where it is impracticable to decrease the measuring radius, and at the same time keep $K_{env,i} > 4$ dB, than it is necessary to increase the equivalent sound absorption area of the room (in short, the room absorption), move the measuring setup to another room, or carry out the measurements in open space.

9.2.2.2 Qualification of the Measuring Environment—the Environmental Correction $K_{env,i}$

The environmental correction $K_{env,i}$ in open space over a sound-reflecting plane equals 0. In rooms, which are limited spaces, the environmental correction $K_{env,i}$ for a hemispherical measurement surface (on the analogy of EN ISO 3744:2010), is calculated for the i-th one-third octave frequency band from the following formula:

$$K_{env,i} = 10 \log_{10}\left(1 + 4S / A_i\right) \text{ dB} \qquad (9.1)$$

where:

    S (m²): the measurement surface area, $S = 2\pi r^2$

    A_i (m²): the room absorption for i-th frequency band

    A_i is calculated from the Sabine's formula:

$$A_i = 0.16 \, (V / T_i) \ \text{m}^2 \tag{9.2}$$

where:

    V (m³): the test room volume

    T_i (s): the room reverberation time for i-th one-third octave frequency band

The environmental correction $K_{\text{env},i}$ for a semi-spherical measurement surface, for the i-th one-third octave frequency band, is calculated from the formula:

$$
\begin{aligned}
K_{\text{env},i} &= 10\log_{10}\left(1 + 4S / A_i\right) \\
&= 10\log_{10}\left(1 + 24.8S \cdot T_i / V\right) \approx 10\log_{10}\left(1 + 25S \cdot T_i / V\right) \ \text{dB}
\end{aligned}
\tag{9.3}
$$

where symbols the same as in Equation 9.2 are used.

The test environment correction in each of the frequency bands in which the sound power level is determined must not exceed 4 dB (i.e., $K_{\text{env},i} \leq 4$ dB for each i).

9.2.2.3 The Choice of Number and Location of Measurement Points on the Measurement Surface

It follows from the performed studies [Mikulski and Radosz 2010; Mikulski et al., 2010] (see also Figure 9.2) and comparison of sound power measurement results obtained for the same sound source but determined based on measurements taken at different number of measurement points, that for 37 or more measurement points, the obtained averaged sound power level value differs from the sound power level value calculated from over 1000 points by less than 1 dB. It can therefore be stated that to the accuracy of 1 dB, the sound power level of the source can be determined based on 37 measurements points. Such a rule of thumb is also consistent with another rule according to which the maximum numerical value of the sound pressure levels difference (expressed in decibels) for any two different points of the measurement surface must be less than the number of measurement points set out in EN ISO 3744:2010 and EN ISO 9295:2015. The number 37 is about four times larger than the provisional quantity of measuring points recommended in EN ISO 3744:2010 and EN ISO 3746:2010 and almost twice as large as the number 20 suggested in EN ISO 9295:2015. From that it follows that in the considered frequency range it is necessary to increase the number of measurement points 2–4 times compared to the number adopted in currently applicable

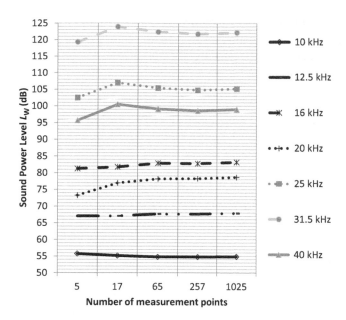

FIGURE 9.2 Results of measurements of the sound power level generated by an ultrasonic washer determined from different numbers of measurement points. (From Mikulski, W. et al., Opracowanie wytycznych akustycznych projektowania obudów dźwiękoizolacyjnych przemysłowych technologicznych źródeł hałasu ultradźwiękowego. Opracowanie modelu obudowy dźwiękoizolacyjnej na wybrane technologiczne źródło hałasu ultradźwiękowego, Warszawa, CIOP-PIB, 2019.)

standards in the frequency range below 10 kHz (and even up to 20 kHz). Note that the number 37 is a result of the assumed even distribution of measurement points on the measurement hemisphere with the pitch of 30° in the polar system of coordinates.

9.2.2.4   Calculation of Average Value of the Sound Pressure Level on the Measurement Surface

The average sound pressure level on the measurement surface is calculated in dB, for the i-th one-third octave frequency band, from the formula:

$$\left\langle L_{p,i} \right\rangle = 10\log_{10}\left(\frac{1}{N} \sum_{n=1}^{N} 10^{0.1 L_{p,n,i}} \right) \text{dB} \qquad (9.4)$$

where:

N: the number of measurement points

$L_{p,n,i}$ (dB): the sound pressure level on the measurement surface at n-th measurement point for the i-th one-third octave frequency band

9.2.2.5   Calculation of the Correction for Background Noise $K_{bgn,i}$

The correction for background noise $K_{bgn,i}$ (in the standard EN ISO 3744:2010, designation $K_{1,i}$ was used for the quantity) for i-th one-third octave frequency band is calculated from the formula:

$$K_{bgn,\,i} = -10\log_{10}\left(1 - 10^{-0.1 \cdot \Delta Lp,\,i}\right) \text{ dB} \tag{9.5}$$

where:

$$\Delta L_{p,\,i} = \left\langle L_{p,\,src,i}\right\rangle - \left\langle L_{p,bgn,i}\right\rangle \text{ dB} \tag{9.6}$$

and

$\langle L_{p,src,i}\rangle$ (dB): the average sound pressure level value on the measurement surface, in i-th one-third octave frequency band, over the period of time in which the source (or the source in enclosure) emits ultrasonic noise (in dB)

$\langle L_{p,bgn,i}\rangle$ (dB): the average sound pressure level value on the measurement surface, in i-th one-third octave frequency band, at the moment when the source does not emit ultrasonic noise, in dB

If $\Delta L_{p,i} > 15$ dB in any frequency band, it can be assumed that the correction $K_{bgn,i}$ equals zero and can be neglected in calculations.

For 3 dB $\leq \Delta L_{p,i} \leq 15$ dB, the correction $K_{bgn,i}$ should be taken into account in calculations.

Values $\Delta L_{p,i} < 3$ dB cannot be accepted. To increase value of the difference, it is necessary to move the measurement surface closer to the sound source (see above), reduce the background noise, or use a sound source with higher sound power level.

9.2.2.6   Calculation of the Correction for Sound Attenuation in Air $K_{air,i}$

The correction resulting from the effect of sound damping in air is determined, for given i-th one-third octave frequency band, from the formula:

$$K_{air,i} = \alpha_{air,i} \cdot r \text{ dB} \tag{9.7}$$

where:

$\alpha_{air,i}$ (dB/m): the coefficient of sound attenuation in air in the i-th one-third octave frequency band (cf. Equation 9.8)

r (m): the measurement surface radius

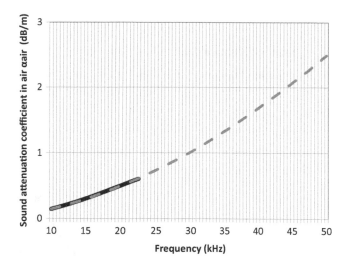

FIGURE 9.3 The coeffcient of sound attenuation in air (heavy line represents the values of EN ISO 9295:2015; dashed line represents the estimation function). (From Mikulski, W., *Arch. Acoust.*, 38, 177–183, 2013; Mikulski, W. et al., Opracowanie wytycznych akustycznych projektowania obudów dźwiękoizolacyjnych przemysłowych technologicznych źródeł hałasu ultradźwiękowego. Opracowanie modelu obudowy dźwiękoizolacyjnej na wybrane technologiczne źródło hałasu ultradźwiękowego, Warszawa, CIOP-PIB, 2019.)

The standard EN ISO 9295:2015 proposes a calculation formula for values of the coefficient of sound attenuation in air α_{air} in the frequency range of 10–22.4 kHz (in napiers per meter, Np/m). Values of the parameter converted into dB/m, for humidity 50% and temperature 22°C are represented in Figure 9.3 in the form of a heavy line. For the frequency range of up to 50 kHz, the values of the coefficient of sound attenuation in air α_{air} were extrapolated with the use of the power function (the dashed line in Figure 9.3). The corresponding regression line is fit with the correlation coefficient $R^2 = 0.9996$.

The coefficient of sound attenuation in air $\alpha_{air,i}$ can be calculated with the use of the following approximate formula:

$$\alpha_{air,i} = 0.0026 \left\{ f_i \right\}^{1.7546} \text{ dB/m} \tag{9.8}$$

where $\{f_i\}$ is the numerical value of the mid-band frequency of i-th one-third octave frequency band expressed in kHz.

Table 9.1 lists and Figure 9.4 depicts values of the correction $K_{air,\,1m}$ for the distance of 1 m in one-third octave bands with mid-band frequencies from the range of 10–40 kHz calculated from the formula of Equation 9.8.

TABLE 9.1

Correction $K_{air, 1m}$ for the Distance 1 m

One-Third Octave Band Mid-Band Frequency (kHz)	10	12.5	16	20	25	31.5	40
Correction $K_{air, 1m}$ (dB)			0.15	0.22	0.34	0.5	0.74	1.11	1.68

Note: Relative humidity 50%, temperature 22°C.

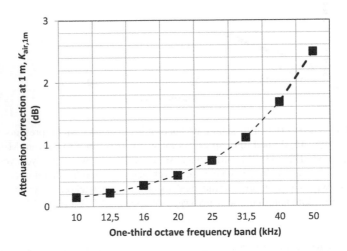

FIGURE 9.4 Correction $K_{air,1 m}$ for the distance of 1 m in one-third octave frequency band (humidity 50%, temperature 22°C).

9.2.2.7 Calculation of the Sound Power Level of Source

The sound power level of a sound source in quasi-free field (including in any room in which the environmental correction $K_{env,i} \leq 4$ dB) over a sound-reflecting surface is determined, for the i-th one-third octave frequency band, from the formula:

$$L_{W,i} = \langle L_{p,i} \rangle + 10 \log_{10}(2\pi r^2) - K_{env,i} - K_{bgn,i} + K_{air,i}(r) \text{ dB} \qquad (9.9)$$

where:

$\langle L_{p,i} \rangle$ (dB): the average sound pressure level on the measurement surface in i-th one-third octave frequency band

r (m): radius of hemispherical measurement surface

$K_{env,i}$ (dB): the environmental correction for the i-th one-third octave frequency band

$K_{bgn,i}$ (dB): the correction for background noise for the i-th one-third octave frequency band

$K_{air,i}(r)$ (dB): the correction for sound attenuation in air for the i-th one-third octave frequency band and distance r

9.3 THE MEASURING METHOD FOR THE SOUND POWER INSULATION OF ENCLOSURES IN THE 10–40 kHz FREQUENCY RANGE

When determining the power insulation value (by taking sound power level of source measurements first without, and next with enclosure) a reference sound source is placed on a sound-reflecting surface, whereas measurements are taken on a hemispherical measurement surface (with radius of, for instance, 1 m) defined over a sound-reflecting ground surface.

The sound power insulation $D_{W,i}$ for an enclosure in one-third octave frequency bands (with mid-band frequencies f_i of 10 kHz, 12.5 kHz, 16 kHz, 20 kHz, 25 kHz, 31.5 kHz, and 40 kHz) is calculated from the formula (the index i denoting individual frequency bands is omitted for brevity):

$$
\begin{aligned}
D_W &= L_{W,\text{w/o}} - L_{W,\text{w}} \\
&= \langle L'_{p,\text{w/o}} \rangle - K_{\text{bgn,w/o}} - K_{\text{env}} + 10\log_{10}\{S\} + K_{\text{air}} \\
&\quad - \left(\langle L'_{p\text{w}} \rangle - K_{\text{bgn,w}} - K_{\text{env}} + 10\log_{10}\{S\} + K_{\text{air}} \right) \\
&= \langle L'_{p,\text{w/o}} \rangle - \langle L'_{p\text{w}} \rangle - K_{\text{bgn,w/o}} + K_{\text{bgn,w}} \quad \text{dB}
\end{aligned}
\tag{9.10}
$$

where:

$L_{W,\text{w/o (w)}}$ (dB): the sound power level determined with the use of the method presented in Section 9.2, for the reference sound source without (w/o) enclosure and the reference sound source with (w) enclosure, respectively

$\langle L'_{p,\text{w/o (w)}} \rangle$ (dB): the average sound pressure level on hemispherical measurement surface determined with the use of the method described in Section 9.2, for the reference sound source without (w/o) enclosure and the reference sound source with (w) enclosure, respectively

$K_{\text{bgn,w/o (w)}}$ (dB): corrections for background noise for the reference sound source without (w/o) enclosure and the reference sound source with (w) enclosure, respectively (as the difference between the sound pressure level on the measurement surface from the reference sound source without enclosure and from the background noise in the analyzed frequency bands, in the majority of cases exceeded 15 dB, then $K_{\text{bgn,w/o}} = 0$)

K_{env} (dB): the environmental correction (despite the fact that the environmental correction $K_{\text{env},i}$ is of no significance from the point of view of determining the sound power insulation characterizing an enclosure (as it is cancelled in calculations), it is still determined for the purpose of the so-called qualification of the measuring environment, i.e., it has to meet the condition $K_{\text{env},i} \leq 4$ dB in any one-third octave frequency band)

S (m²): the measurement surface area

K_{air} (dB): the correction for sound attenuation in air (also cancels in calculations)

As it was already mentioned, Equation 9.10 applies to each one-third octave frequency band with given mid-band frequency.

The number of measurement points used to determine the sound insulation of a sound-insulating enclosure must not be less than 37 (on the analogy of determination of the sound power level; cf. Section 9.2.2). For that number, measurement points may be distributed every 30° in both horizontal and vertical planes (Figure 9.5).

The average sound pressure level on the measurement surface (for the reference sound source both without and with enclosure) is calculated in dB, for the one-third octave frequency band with mid-band frequency f_i, from the formula of Equation 9.4.

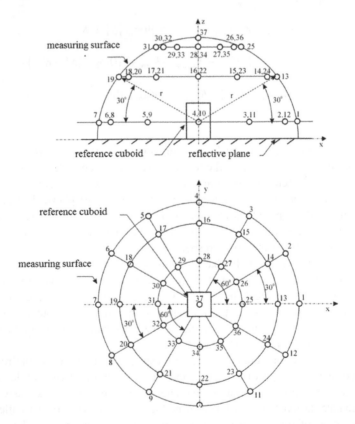

FIGURE 9.5 Location of 37 measurement points on hemispherical measurement surface. (From Mikulski, W. et al., Opracowanie wytycznych akustycznych projektowania obudów dźwiękoizolacyjnych przemysłowych technologicznych źródeł hałasu ultradźwiękowego. Opracowanie modelu obudowy dźwiękoizolacyjnej na wybrane technologiczne źródło hałasu ultradźwiękowego, Warszawa, CIOP-PIB, 2019.)

The correction for background noise $K_{bgn,i}$ for the i-th one-third octave frequency band (for the source both without and with enclosure) is calculated in dB from the formula of Equation 9.5.

When the sound power insulation value for a sound-insulating enclosure is calculated, it is possible, in the majority of cases, to omit the correction for background noise in measurements concerning the reference sound source without enclosure. This follows from the fact that if the measured sound pressure levels from the reference sound source in the tested enclosure are higher than the background noise levels, then in the course of measurements for the source without enclosure, the measured power levels will be significantly higher (by the enclosure sound insulation value).

In Equation 9.10, the environmental correction $K_{env,i}$ is eliminated in the course of calculations. However, to ensure that the measurements are carried out in conditions in which the sound energy coming directly from the sound source exceeds significantly the energy of reflected waves, values of the correction, as in Section 9.2.2, should be less than 4 dB ($K_{env,i} \leq 4$ dB). The environmental correction $K_{env,i}$ in rooms for the hemispherical measurement surface is calculated, in dB, for the mid-band frequency f_i, from Equations 9.1 and 9.3.

9.4 RESULTS OF EXAMINATION OF THE SOUND POWER INSULATION OF RIGID-FRAME PLYWOOD ENCLOSURES WITH AND WITHOUT SOUND-ABSORBING MATERIAL INSIDE

The setup at which the sound power insulation was determined for sound-insulating enclosures with the use of the measuring-computational method in the 10–40 kHz frequency range, included:

- the testing environment (a room meeting specific criteria),
- the sound-reflecting surface on which the reference sound source and the sound-insulating enclosure are placed,
- the reference (laboratory) sound source, and
- the measuring apparatus—a standard apparatus for noise spectrum measurements in one-third octave frequency bands with mid-band frequencies from the range 10–40 kHz, for example, SVAN 912 or SVAN 979 (Svantek) or PULS (Brüel & Kjær).

The testing environment is a room (quasi-anechoic chamber) with a sound-reflecting substrate. The reflecting surface is a smooth terracotta slab with dimensions 4 m × 4 m. The environment meets the requirements which are established for the environmental correction $K_{env,i}$ (cf. Section 9.3). The requirements concern the room absorption and the measurement radius r (1 m). Figure 9.6 is a graph representing results of measurements of the environmental correction $K_{env,i}$ in the laboratory room of the Central Institute for Labour Protection–National Research Institute in

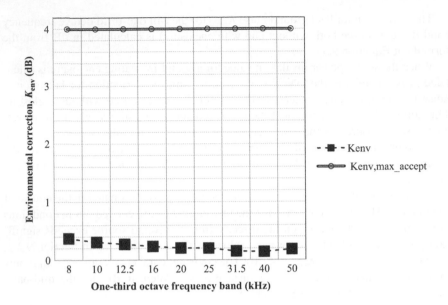

FIGURE 9.6 Environmental correction $K_{env,i}$ for the assumed measuring radius $r = 1$ m in one-third octave frequency band.

Warsaw in which sound power insulation tests and power level measurements for noise sources in the 10–40 kHz frequency range were carried out. In the room, values of the environmental correction $K_{env,i}$ in the entire 10–40 kHz frequency range are significantly less than the maximum acceptable limits (K_{env,i,max_accept}).

The reference sound source used in studies on the sound power insulation characterizing sound-insulating enclosures in the frequency range of about 8–50 kHz (developed and constructed in the above-mentioned laboratory) is shown in Figure 9.7, whereas Figure 9.8 presents results of the source sound power level measurements for electric power supply values of 4 W and 16 W.

Directivity patterns for the sound emission from the reference source are shown in Figure 9.9—for directivity pattern in horizontal plane and Figure 9.10—for directivity pattern in vertical plane.

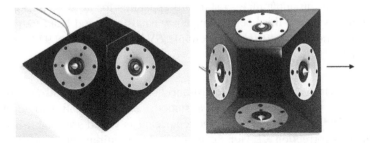

FIGURE 9.7 The reference sound source in the frequency range of about 8–50 kHz (a column loudspeaker). The angle between the base and side planes of the source is 45° (the arrow marks the source main axis parallel to the ground plane).

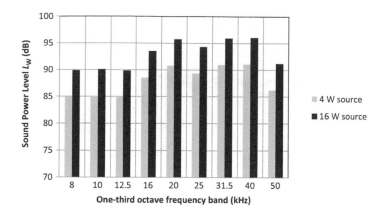

FIGURE 9.8   The sound power level of the reference sound source in one-third octave frequency bands.

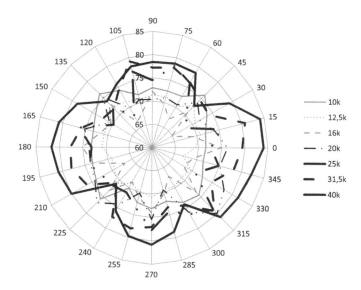

FIGURE 9.9   The reference sound source radiation directivity in horizontal plane (numbers in the column next to vertical axis of the graph are sound pressure level values in dB, whereas numbers on circumference of the polar chart are values (in degrees) of the plane angle from the sound source main axis marked with arrow in Figure 9.7).

The maximum difference between sound pressure levels for each of the one-third octave frequency bands on hemispherical measurement surface (i.e., from the values measured at 37 measurement points) is included in the range of 6–22 dB.

The sound source emits the energy with maximum intensity at the angle of 45° relative to the ground plane (i.e., in directions corresponding to principal axes of loudspeakers).

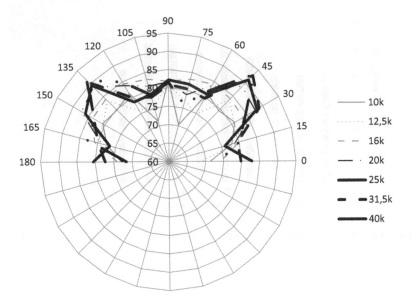

FIGURE 9.10 The reference sound source radiation directivity in vertical plane (numbers in the column next to vertical axis of the graph are sound pressure level values in dB, whereas numbers on circumference of the polar chart are values (in degrees) of the plane angle from the sound source main axis marked with arrow in Figure 9.7).

The examined rigid-frame enclosure had the form of a skeleton made of aluminum angle bars lined from the inside with a foam to seal junctions of the frame to which external walls made of plywood were fixed and tightly pressed to form a seal. The wall joints were additionally sealed with a sound-absorbing material used typically to manufacture molded hearing protectors. The frame had the form of a cube with a 400-mm long edge (Figure 9.11).

Figure 9.12 shows a plot of standard deviation values for the sound pressure level on the hemispherical measurement surface, for the reference sound source in the sound-insulating enclosure (without sound-absorbing material inside the enclosure), and for the reference sound source without enclosure. Dispersion of sound pressure level values is higher for the reference sound source without enclosure compared to the reference sound source with enclosure (which means that the enclosure makes the sound energy propagation into the environment less directive).

Figure 9.13 is a graph showing results of sound power insulation measurements for a rigid-frame enclosure in the form of a regular cube with walls made of 6- and 12-mm thick plywood. Results of the sound power insulation measurements are as follows: for the frequency of 10 kHz, 25.4 dB and 31.4 dB (for the 6- and 12-mm thick enclosure, respectively), and for the frequency 40 kHz, 42.0 dB and 46.2 dB (for the 6- and 12-mm thick enclosure, respectively).

Figure 9.14 shows results of measurements of the sound power insulation for a rigid-frame enclosure in the form of a regular cube with walls made of 6-mm thick plywood: without lining (as in Figure 9.14) and with a sound-absorbing material

FIGURE 9.11   The frame of the enclosure used in the study on sound power insulation of rigid-frame enclosures or their walls in the 10–40 kHz frequency range.

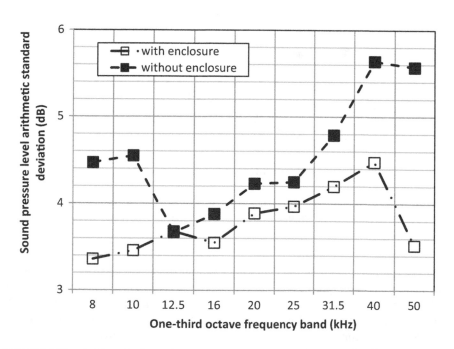

FIGURE 9.12   Arithmetical standard deviation of sound pressure level values on the measurement surface for the reference sound source with enclosure and for the reference sound source without enclosure in one-third octave frequency band.

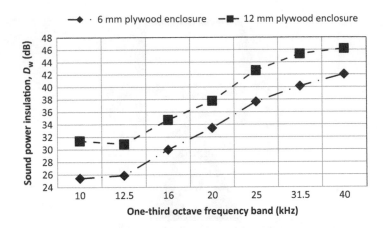

FIGURE 9.13   The sound power insulation $D_{W,i}$ of a rigid-frame enclosure with walls made of 6 and 12-mm thick plywood in one-third octave frequency band.

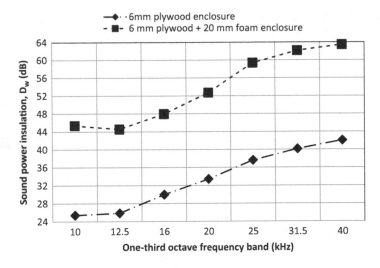

FIGURE 9.14   The sound power insulation $D_{W,i}$ of a rigid-frame enclosure with walls made of 6-mm thick plywood, with and without 20-mm thick polyurethane sheet in one-third octave frequency band.

covering the inner walls of the enclosure (20-mm thick polyurethane foam sheet). The measured sound power insulation values were as follows: at the frequency of 10 kHz, 25.4 dB and 45.4 dB for 6-mm enclosure and 6-mm enclosure with 20-mm foam sheet, respectively, and at the frequency 40 kHz, 42.0 dB and 63.5 dB for 6-mm enclosure and 6-mm enclosure with 20-mm foam sheet, respectively.

Values of the sound power insulation characterizing the tested plywood enclosures increased when the frequency was doubled: for 6-mm walls by about 9.6 dB, for 12-mm walls by about 9.3 dB, and for 6-mm walls with 20-mm foam sheet on inner walls by about 11.8 dB.

The average difference between the sound power insulation values for the enclosure with walls made of 12-mm and 6-mm thick plywood is 4.9 dB, and the difference for the enclosure with 6 mm thick walls lined with 20 mm foam inside and 6-mm thick enclosure without foam lining is 20.1 dB.

The above-quoted results indicate that the effect of providing a lining of a sound-absorbing material on inner walls of the enclosure (about 20 dB) was significantly stronger than the result of doubling the thickness of plywood enclosures (about 5 dB) in the 10–40 kHz noise frequency range.

9.5   SUMMARY

When reduction of exposure to noise is vital, one of the most effective measures to be implemented in the work environment are sound-insulating enclosures designed to be put over a sound source. In industrial plants and service businesses, there are many noise sources emitting the sound energy in the 10–40 kHz frequency range (e.g., ultrasonic washers, ultrasonic welders, ultrasonic erosion machines). Such machines and devices are frequently the sources of immission of noise harmful to humans in that very frequency range. Therefore, sound-insulating enclosures should be used wherever possible to reduce the energy radiated to the environment by sources of ultrasonic noise. In view of the present state of the art and absence of internationally agreed standards, there are currently no commonly accepted measuring methods which could be used to assess the acoustic effectiveness of sound-insulating enclosures in the 10–40 kHz frequency range. This chapter presented a measuring method applicable for determination of the sound power insulation characterizing enclosures and a sound power level of source measuring method adapted to that specific sound frequency range. The measuring method for the sound insulation of enclosures is based on determination of the insertion loss calculated as the difference between the sound power of a reference sound source without and with enclosure. For that reason, the method allows for omission of effects connected with sound attenuation in air and the effect of acoustic properties of the room in which measurements are taken (provided that sufficient ratio is ensured of the sound pressure level corresponding to the direct sound to the sound pressure level of reflected sound and background noise pressure level). The minimum number of measurement points following from directional nature of radiation generated by sound sources in that frequency range was assumed to be about 37.

It was found that the sound power insulation of a rigid-frame enclosure made of plywood sheeting with the thickness of 6 mm, in the 10–40 kHz frequency range, falls into the range of 25–42 dB (it increases monotonically at the average rate of about 9.6 dB per doubled frequency). Providing the inner walls of the enclosure with a 20-mm thick polyurethane foam sheet lining increases its sound insulation by about 20.1 dB on average in the analyzed frequency range. After doubling the thickness of walls of an enclosure without sound-absorbing material inside, the sound power insulation of the enclosure turned out to be only about 4.9 dB higher in the examined frequency range.

That proves unambiguously that sound-insulating enclosures with thin walls and with sound-absorbing material inside show better sound insulation than enclosures with much thicker walls but without sound-absorbing material inside.

ACKNOWLEDGMENTS

The author wishes to express deep thankfulness to his colleagues from the Central Institute for Labour Protection–National Research Institute for their assistance in taking measurements in the years 2010, 2017–2019 [Mikulski et al. 2010, 2019], especially to Jerzy Kozlowski for his permission to publish Figures 9.1 and 9.5.

REFERENCES

Directive 2000/14/EC of the European Parliament and of the Council of May 8, 2000 on the approximation of the laws of the Member States relating to the noise emission in the environment by equipment for use outdoors. *OJ L 162*, July 3, 2000, pp. 1–78.

Directive 2003/88/EC of the European Parliament and of the Council of November 4, 2003 concerning certain aspects of the organization of working time. *OJ L 299*, November 18, 2003, pp. 9–19.

Directive 2006/42/EC of the European Parliament and of the Council of May 17, 2006 on machinery, and amending Directive 95/16/EC. *OJ L 157*, June 9, 2006, pp. 24–86.

Dobrucki A., B. Żółtogórski, P. Pruchnicki, and R. Bolejko. 2010. Sound-absorbing and insulating enclosures for ultrasonic noise. *Arch Acoust* 35(2):157–164. DOI:10.2478/v10168-010-0014-4.

EN ISO 3744:2010. 2010. Acoustics—Determination of sound power levels and sound energy levels of noise sources using sound pressure—Engineering methods for an essentially free field over a reflecting plane. Brussels, Belgium: European Committee for Standardization.

EN ISO 3746:2010. 2010. Acoustics—Determination of sound power levels and sound energy levels of noise sources using sound pressure—Survey method using an enveloping measurement surface over a reflecting plane. Brussels, Belgium: European Committee for Standardization.

EN ISO 9295:2015. 2015. 2010. Acoustics—Determination of high-frequency sound power levels emitted by machinery and equipment. Brussels, Belgium: European Committee for Standardization.

EN ISO 11546-2:2009. 2009. Acoustics—Determination of sound insulation performances of enclosures—Part 2—Measurements in situ (for acceptance and verification purposes). Brussels, Belgium: European Committee for Standardization.

EN ISO 15667:2000. 2000. Acoustics—Guidelines for noise control using enclosures and cabins. Brussels, Belgium: European Committee for Standardization.

Engel, Z., and J. Sikora. 1998. *Obudowy Dźwiękochłonno-Izolacyjne: Podstawy Projektowania i Stosowania.* [Sound absorbing-insulating enclosures. Foundations of design and application]. Kraków: Wydawnictwa AGH.

Grzesik, J., and E. Pluta. 1983. High-frequency hearing risk of operators of industrial ultrasonic devices. *Int Arch Occup Environ Health* 53(1):77–88.

Holmberg, K., U. Landstrom, and B. Nordstom. 1995. Annoyance and discomfort during exposure to high-frequency noise from an ultrasonic washer. *Percept Mot Skills* 81(3):819–827.

ISO 9613-1:1993. 1993. Acoustics—Attenuation of sound during propagation outdoors—Part 1—Calculation of the absorption of sound by the atmosphere. Geneva, Switzerland: International Organization for Standardization.

ISO 9613-2:1996. 1996. Acoustics—Attenuation of sound during propagation outdoors—Part 2: General method of calculation. Geneva, Switzerland: International Organization for Standardization.

Kling, C., C. Koch, and R. Kühler. 2015. Measurement and assessment of airborne ultrasound noise. *Proceedings of The International Congress on Sound and Vibration, ICSV 22*, July 12–16, 2015, Florence, Italy.

Lawton, B.W. 2013. Exposure limits for airborne sound of very high frequency and ultrasonic frequency. ISVR Technical Report No: 334/2013. Southampton: University of Southampton.

Mikulski, W. 2013. Method of determining the sound absorbing coefficient of materials within the frequency range of 5,000–50,000 Hz in a test chamber of a volume of about 2 m³. *Arch Acoust* 38(2):177–183. DOI:10.2478/aoa-2013-0020.

Mikulski, W. 2019. The effect of the application sound-absorbing material inside enclosure on the sound insulation of enclosure in the frequency range 10–40 kHz. *Proceedings of the Conference Noise Control 2019*, May 26–29, 2019, Warszawa-Janów Podlaski, Poland.

Mikulski, W., and J. Radosz. 2010. The measurements of ultrasonic noise sources in the frequency range from 10 to 40 kHz. *Proceedings of The 17th International Congress on Sound and Vibration*, July 18–22, 2010, Cairo, Egypt.

Mikulski, W., J. Radosz, and J. Kozłowski. 2010. Wykonanie prototypu źródła akustycznego emitującego hałas ultradźwiękowy o stabilnej i dużej mocy akustycznej. Badania weryfikacyjne prototypu źródła hałasu ultradźwiękowego. Opracowanie stanowiska badawczego do badań emisji, propagacji oraz imisji hałasu w zakresie częstotliwości od 10 kHz do 50 kHz. [Making a prototype of acoustic source emitting ultrasonic noise of stable and high acoustic power. Verification tests of the ultrasonic noise source. Development of a setup for research on emission, propagation, and immission of noise in the frequency range from 10 kHz to 50 kHz]. Warszawa: CIOP-PIB.

Mikulski, W., J. Radosz, D. Pleban, A. Budziak, and J. Kozłowski. 2019. Opracowanie wytycznych akustycznych projektowania obudów dźwiękoizolacyjnych przemysłowych technologicznych źródeł hałasu ultradźwiękowego. Opracowanie modelu obudowy dźwiękoizolacyjnej na wybrane technologiczne źródło hałasu ultradźwiękowego. [Research on acoustic emission of technological directive sources of high-acoustic-power ultrasonic noise and studies on acoustic effectiveness of sound-insulating enclosures for such sources. Development of a model of the sound-insulating enclosure for a selected technological ultrasonic noise source]. Warszawa: CIOP-PIB.

Pawlaczyk-Łuszczyńska, M., A. Dudarewicz, and M. Śliwińska-Kowalska. 2007. Źródła ekspozycji zawodowej na hałas ultradźwiękowy – Ocena wybranych urządzeń. [Sources of occupational exposure to ultrasonic noise—Evaluation of selected devices]. *Med Pr* 58(2):105–116.

Pleban, D. 2013. Method of testing of sound absorption properties of materials intended for ultrasonic noise protection. *Arch Acoust* 38(2):191–195. DOI:10.2478/aoa-2013-0022.

Pleban, D., and W. Mikulski. 2015. Determination of sound insulation properties of barriers for ultrasonic noise reduction. *Proceedings of the 22th International Congress on Sound and Vibration*, July 12–16, 2015, Florence, Italy.

Pleban D., and W. Mikulski. 2018. Methods of testing of sound insulation properties of barriers intended for high frequency noise and ultrasonic noise protection. *Strojnícky Casopis-J Mech Eng* 68(4):55–64.

Pleban, D., J. Radosz, and B. Smagowska. 2018. Occupational risk assessment related to ultrasonic noise. *Proceedings of the 47th International Congress and Exposition on Noise Control Engineering INTER-NOISE 2018*, August 26–29, 2018, Chicago.

Radosz, J. 2014. Uncertainty due to instrumentation for sound pressure level measurement in high frequency range. *Noise Control Eng J* 62(4):186–195. DOI:10.3397/1/376219.

Radosz, J., and D. Pleban. 2018. Ultrasonic noise measurements in the work environment. *J Acoust Soc Am* 144(4):2532–2538. DOI:10.1121/1.5063812.

Smagowska, B. 2013a. An objective and subjective study of noise exposure with in the frequency range from 10 to 40 kHz. *Arch Acoust* 38(4):559–563. DOI:10.2478/aoa-2013-0066.

Smagowska B. 2013b. Ultrasonic noise sources in a work environment. *Arch Acoust* 38(2):169–176. DOI:10.2478/aoa-2013-0019.

Smagowska, B. and M. Pawlaczyk-Łuszczyńska. 2013. Effects of action of ultrasonic noise on the human body – A bibliographic review. *Int J Occup Saf Ergon* 19(2):195–202. DOI:10.1080/10803548.2013.11076978.

Śliwiński, A. 2013. Assessment of ultrasonic noise hazard in work places environment. *Arch Acoust* 38(2):243–252. DOI:10.2478/aoa-2013-0029.

Śliwiński, A. 2016. On the noise hazard assessment within the intermediate range of the high audible and the low ultrasonic frequencies. *Arch Acoust* 41(2):331–338. DOI:10.1515/aoa-2016-0034.

10 Studies on Sound Insulation Effectiveness of Phononic Crystals

Jan Radosz

CONTENTS

10.1 INTRODUCTION

By definition, phononic crystals are periodic solid inclusions in a fluid medium [Khelif and Adibi 2016]. In practice, the structures are constructed of evenly distributed scatterers. The feature of such systems most important from the acoustical point of view is the effect consisting in occurrence of the so-called forbidden bands or energy gaps—definite frequency ranges in which the sound wave cannot propagate in the structure. The size and location of forbidden bands or band gaps in the spectrum depend on many factors such as the wave incidence angle or spatial distribution of scatterers.

The elementary component of an ideal atomic crystalline structure can be defined as a specific (as far as their arrangement and orientation is concerned) group of atoms in space [Kittel 2004]. When such an elementary component is repeated infinitely many times and located at determined points of space called *nodes*, then a crystalline structure is formed. In the simplified two-dimensional case, a lattice can be defined by two translation vectors $\underline{a}_1$ and $\underline{a}_2$ in such way that the arrangement of atoms in the crystalline structure looks the same when seen from either point $\underline{r}$ or from any other point $\underline{r}'$ which is obtained by moving $\underline{r}$ with any translation vector $\underline{T}$

$$\underline{r}' = \underline{r} + \underline{T} \tag{10.1}$$

where $\underline{T} = u_1 \cdot \underline{a}_1 + u_2 \cdot \underline{a}_2$ and u_1, u_2 are any two integers.

It can therefore be stated that the crystalline structure is invariable, translation-independent, and in some cases also rotation-independent. Crystalline structures can be defined in vector spaces of different dimensions, as one-dimensional (1D), two-dimensional (2D), and three-dimensional (3D) systems. The elementary cell is a cell corresponding to a single lattice node of a structure with translational symmetry. A crystalline lattice can therefore be characterized based on geometry of its elementary cell.

With the use of group theory it has been proven that there exists only one unique one-dimensional crystallographic system, five two-dimensional systems, and fourteen three-dimensional crystallographic systems. In the present study, only 2D system are discussed for which five different Bravais translational lattices can be defined (Figure 10.1).

The polygons based on vectors $\underline{a}_i$ define the elementary cell constituting the smallest possible geometrical figure which can be used as a base for a given crystalline structure. Translation of the elementary cell by the vector $\underline{T}$ defined in space generates a periodic lattice in space.

The elementary cell contains the whole of crystallographic information concerning the lattice structure, whereas the crystalline lattice constants are the distances between centers of neighboring cells in the lattice equaling $a_i = |\underline{a}_i|$ The fraction of lattice nodes on the elementary cell surface is known as the *packing fraction*, a quantity important from the point of view of the phononic crystal optimization problem.

The science of crystallography introduces the concept of the reciprocal lattice, which is not defined in the actual physical space of positions in which a crystal can be seen. The reciprocal lattice cannot therefore be associated with any material objects such as atoms. Apart from that, the reciprocal lattice has all other attributes of crystalline lattice—it is a structure in the form of a lattice of nodes (specifically defined points in space) and its symmetry can be analyzed with the use of the same methods that apply to crystalline lattices of atoms. The elementary cell of the reciprocal

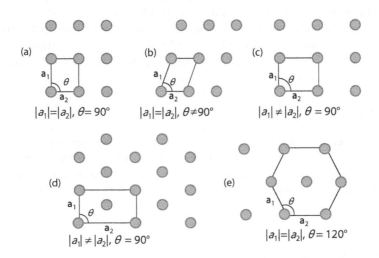

FIGURE 10.1 Five basic 2D Bravais lattices: (a) square; (b) oblique; (c) rectangular; (d) centered rectangular; and (e) hexagonal. (From Chong, Y.B., Sonic crystal noise barriers, PhD Thesis, The Open University, http://oro.open.ac.uk/44502/, accessed January 28, 2020, 2012.)

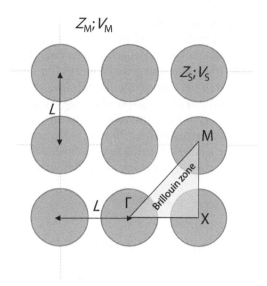

FIGURE 10.2 A schematic representation of square lattice of cylindrical scatterers with Z_s acoustic impedance (scatterer), V_s sound velocity (scatterer), Z_m acoustic impedance (medium), V_m sound velocity (medium), and L lattice constant (distance between scatterers).

lattice with the least possible volume is called the *first Brillouin zone*. A Brillouin zone is the least space limited by planes orthogonal to lattice-defining vectors and crossing them at their half length (Figure 10.2).

Diffraction of waves on crystals is closely related to Brillouin zones [Brillouin 1946]. Each wave with the wave vector drawn from the center of the first Brillouin zone to its boundary is a source of a diffracted wave.

Another physical property of a crystalline lattice is characterized by the packing factor for a given lattice constant and dimensions of scatterers. The packing factor *PF* is defined as the ratio of volume occupied by scatterers to the whole volume occupied by the elementary cell. In case of cylindrical scatterers with radius r_0, the packing factor values for square and hexagonal lattices can be calculated from:

$$PF_{sqr} = \frac{\pi r_0^2}{a^2} \tag{10.2}$$

and

$$PF_{hex} = \frac{2\pi r_0^2}{a^2} \tag{10.3}$$

respectively. Properties of phononic crystals can be interpreted with the use of the same methods which were developed in the theory of semiconductor materials. The energy gap observed in semiconductors is a range of energy of free electrons present in a solid state body, in which electrons are intensively scattered on atoms.

As a result, the system contains no electrons with energy from that specific range. Properties of energy gaps are determined by size, periodicity, and arrangement of atoms in the crystalline lattice of semiconductors on which diffraction of electrons occurs according to the Bloch theorem [Khelif and Adibi 2016]. The same wave reflection model expressed in the form of the Bragg's law formulated originally for the diffraction of light, can be applied by way of analogy to sound waves. A portion of the wave incident in the crystalline structure will be reflected by the first layer of scatterers, whereas the remaining portion will pass to the second layer when a portion of which will again be let through to the next layers and the remainder reflected.

Bragg's law can be expressed with the use of the formula:

$$2L\sin\theta = n\lambda \tag{10.4}$$

where:
L(m): the distance between planes (lattice constant)
λ(m): the wavelength
θ: the angle between the incident ray and the scattering plane
n: any integer

A phononic crystal can be defined as a finite system of scatterers situated in a homogenous medium. The geometry of a phononic crystal is determined by the form of a single elementary cell making up the Bravais lattice. In such periodic structures, interference phenomena based on diffraction are the rationale behind occurrence of entirely forbidden frequency bands. Figure 10.2 shows a square lattice of cylindrical scatterers immersed in a homogenous medium and sound wave propagation directions defining the first Brillouin zone.

The fundamental Bragg's frequencies occur where crystalline lattice constant L and $\sqrt{2}L$ in directions ΓX and ΓM equals half the wavelength λ. Those frequencies are determined from the following formulae:

$$f_{\text{Bragg}\Gamma\text{X}} = \frac{v_{\text{m}}}{2L} \text{ Hz} \tag{10.5}$$

and

$$f_{\text{Bragg}\Gamma\text{M}} = \frac{v_{\text{m}}}{2\sqrt{2}L} \text{ Hz} \tag{10.6}$$

where:
v_{m}(m/s): the sound velocity in the medium
L(m): the lattice constant (distance between scatterers)

A complete band gap occurs when Bragg's resonance bands are wide enough to overlap. Width of each of the Bragg's resonances depends on the difference in the acoustic impedance between scatterers and the medium and on the packing factor (a function of the lattice constant). The packing factor, on one hand, must be large enough to reduce the sound transmission through the structure around scatterers, and on the other, it must not be too large so as to avoid the risk of undesirable constructive interference.

To occur, the band gap phenomenon requires that there is a difference between physical properties such as the specific density and the sound velocity between the scatterer and the medium. Authors of certain scientific papers suggest that the nature of the medium should be a base for making a distinction between phononic crystals and the so-called sonic crystals [Caballero et al. 2000; Liu et al. 2000; Sanchis et al. 2001; Hu et al. 2005; Chong 2012; Guild et al. 2015; Gupta et al. 2015; Morandi et al. 2015; Peiro-Torres et al. 2016]. If the medium is a solid state matter, then the term "phononic crystal" is used. In a phononic crystal, in view of nature of the medium, both longitudinal waves and transverse waves can be observed. On the other hand, no transverse waves are considered in the case of sonic crystals. Scatterers are made typically of solid materials, while the medium is usually the air or a liquid, so a high contrast in the acoustic impedance value between them is obtained. However, no consistent approach to the terminology of the subject has been worked out in the literature to date [Khelif et al. 2003, 2006; Khelif and Adibi 2016; Goffaux and Vigneron 2001; Goffaux et al. 2003; Oudich et al. 2010]. For the purpose of this chapter, the term "phononic crystal" will be used regardless of the type of medium in which scatterers are placed.

Recent years see an increasing interest in the possibility of using phononic crystals as sound barriers [Chen and Ye 2001; Sánchez-Pérez et al. 2002; van der Aa and Forssén 2014; Morandi et al. 2016; Peiro-Torres et al. 2016], particularly in view of reports on studies in which sound attenuation as high as 25 dB was obtained [Sánchez-Pérez et al. 1998, 2002]. However, barriers in the form of "conventional" phononic crystalline structures—such as cylindrical scatterers in square lattice—offer significant attenuation only in narrow frequency bands and therefore are ineffective in the case of wide-band sound fields. To address the issue, the study presented in this chapter was aimed at extending the range of noise control options by application of phononic crystalline structures and optimizing them for selected types of noise.

Application of phononic crystals to noise suppression is a relatively young and still developing scientific discipline [Pennec et al. 2010; Gupta 2014]. Most of the research effort in this area concerns theoretical considerations and computer-based modeling in search of optimal crystalline structures. Development of technical solutions (physical models) including the stage of model optimization with the use of acoustic cameras or intensity probes can be decisive in determining the necessity to take into account in theoretical considerations concerning factors which were not taken into account to date.

The first in-depth work on wave propagation in periodic structures was published in 1946 [Brillouin 1946]. In his book, Brillouin discussed the mathematical foundations of various problems concerning, among other topics, the solid state theory.

The search for acoustic structures with energy gaps started with theoretical papers published by Sigalas and Economou [1992, 1993] who proved that the structure composed of a periodic three-dimensional lattice of identical spheres with high density immersed in a medium with low density resulted in occurrence of gaps in the sound frequency spectrum. Since then, a plurality of papers on the subject have been published. Among early experimental works on sound in periodic structures it is worth mentioning a report on studies concerning attenuation of sound observed in a minimalist sculpture by Eusebio Sempere [Martínez-Sala et al. 1995]. The sculpture is constructed of cylindrical steel bars with the diameter of 29 mm arranged in a square lattice with the lattice constant of 100 mm. On-site tests confirmed significant attenuation of sound for the frequency of about 1670 Hz within the structure. The capability of phononic crystals to suppress road communication, air traffic, or industrial noise has been demonstrated in a field experiment with the use of PVC tubes with the outer diameter of 160 mm arranged in a oblique lattice [Sánchez-Perez et al. 2002], for which the average sound attenuation of 10–20 dB for frequencies from 1000 to 4000 Hz was obtained. In other tests, a significant energy gap was observed in a rectangular lattice constructed of hollow cylindrical copper scatterers (diameter 28 mm, wall thickness 1 mm) [Vasseur et al. 2002]. In the case of waves propagating perpendicularly to the lattice plane (direction [1 0 0]), the band gap occurs for frequencies from 4000 to 6800 Hz. In other experiments, strong attenuation of sound waves was obtained with the use of periodic structures of steel bars [Sánchez-Perez et al. 1998], hollow copper cylinders [Vasseur et al. 2002], aluminum cylinders with two different diameters [Caballero et al. 2000], and wooden cylinders [Sánchez-Perez et al. 1998; Rubio et al. 1999]. Miyashita [2002] tested phononic crystals made of acrylic rods (with diameter of 20.4 mm) arranged in a square lattice (lattice constant 24 mm) in air and observed occurrence of a band gap between 6800 and 9500 Hz (attenuation up to 25 dB). In another study, hollow steel square sections (length 500 mm, side width 30 mm, wall thickness 3 mm) were arranged to form a square lattice (lattice spacing of 42.5 mm) and rotated by the angle of 40° relative to their axes [Goffaux and Sánchez-Dehesa 2003]. Sound attenuation of up to 30 dB in the range 2000–6000 Hz was obtained in that case.

The scope of scientific research on the subject has also included studies on three-dimensional structures of phononic crystals [Sigalas and Economou 1993; Liu et al. 2002; Sainidou et al. 2002; Chandra et al. 2004; Larabi et al. 2007; Oudich et al. 2010; Lardeau et al. 2016], application of numerical methods to analysis of phononic crystals [Laude et al. 2009; Romero-Garcia et al. 2009; Shakouri et al. 2017], waveguiding [Garcia et al. 2003; Hirsekorn et al. 2004; Phani et al. 2006; Hsiao et al. 2007], and studies on surface sound waves propagating within two-dimensional phononic structures [Lu et al. 2017].

Literature of the subject includes indications that from the acoustical point of view, occurrence of the band gaps phenomenon alone is insufficient for effective application of phononic crystals as acoustic barriers (or similar anti-noise protection measures) [Sánchez-Perez et al. 2002; Garcia et al. 2003; Tang 2018]. Both size (width) and position of band gaps in the sound spectrum depend on many factors such as the wave incidence angle or arrangement of scatterers. In recent years, numerous variants of crystalline-like structures were examined, including quasi-crystals with the intent to improve their insulating power, but a real turning point in the art was a paper [Liu et al. 2000] in which the authors proposed to use scatterers with additional local properties (e.g., sound-absorption) which significantly reduces the dependence of attenuation on the wave incidence angle and increases both the effectiveness of damping and the width of the forbidden frequency band. Such an approach offers new opportunities in the area of research on development of effective noise barriers.

Recent scientific reports available in the world literature on new noise control methods concern, among other things, improvement and standardization of methods used

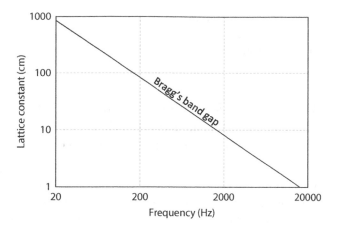

FIGURE 10.3 Occurrence of the band gap in conventional phononic crystals depending on the crystalline lattice constant.

to assess effectiveness of acoustic crystalline structures [Morandi et al. 2015], application of phononic crystals as windows [Lee et al. 2017a, 2017b; Tang 2018], noise suppression with the use of membrane-like structures with the use of the movable impedance approach, application of radial sonic crystal structures [Gupta et al. 2015], application of two-dimensional phononic crystals in heat exchanger structures aimed at reduction of sound transmission and energy recovery [Gonella et al. 2009].

A separate class of literature still pertains to theoretical considerations and numerical modeling aimed at expansion of the range of available noise control options by using crystalline structures [Cervera et al. 2001; Castiñeira-Ibáñez et al. 2010; Godinho et al. 2016; Lardeau et al. 2016; Montiel et al. 2017].

The theory and practice of phononic crystals is still a relatively young scientific discipline and numerous important issues concerning the topic remain unsolved. Conventional phononic crystals designed for low frequencies require large distances between scatterers (Figure 10.3). In the area of new noise control technical solutions, development of structures with the sound attenuation frequency range as wide as possible is an issue of special interest.

10.2   EFFECTIVENESS OF INSULATION FROM AIRBORNE SOUNDS

Analysis of sources available in the literature concerning the subject suggests the need to establish standardized methods for assessment of acoustic insulating power of barriers based on sonic crystals. Typically, relevant measurements are carried out in anechoic chambers [Martinez-Sala et al. 1995; Sánchez-Perez et al. 1998, 2002]. Certain studies [Morandi et al. 2015] deal with verification of measuring apparatus configuration recommended to be used for sound insulation measurements in CEN standards. Availability of standardized methods enables direct comparisons of sound insulating properties demonstrated by sound barriers. However, recent studies [Morandi et al. 2016; Radosz 2019] revealed that the use of standardized measurement methods intended for classic sound screens where selection of a relatively small number of measurement points is recommended might prove to be insufficient for the purpose of comprehensive assessment of effectiveness of sonic crystal-based barriers.

10.3   EXAMINATION OF SOUND FIELD AROUND PHONONIC CRYSTALLINE STRUCTURES WITH THE USE OF ACOUSTIC PARTICLE VELOCITY MEASUREMENTS

To examine thoroughly the sound field around phononic crystalline structures, the method known as "Scan & Paint" was used which is based on measuring the acoustic particle velocity. In the system, the measuring probe is moved manually over a selected measurement plane, while a video camera is used to register the measurement process. The data recorded by the camera is used in the final processing stage to identify the position of the probe for each frame of the video material. The registered signals are divided into many segments with the use of a spatial discretization algorithm assigning a position in space to each of them. Variations in the acoustic field can be then calculated with the use of a separate mesh of values and application of linear interpolation between the mesh nodes. The results are ultimately matched with the measuring environment image in order to obtain a visual representation.

To visualize the sound field, the Microflown Scan & Paint 3D system was used. The phase shift between the acoustic pressure and the acoustic particle velocity is the quantity which depends on the sound propagation direction and distance from the sound source. That means that at any instant of time, the acoustic pressure and each of orthogonal components of the acoustic particle velocity can be combined into a complex vector. As the acoustic pressure and the acoustic particle velocity are measured at the same time, calculations of the sound intensity in three-dimensional space can be performed directly, without the aid of approximations. The sound intensity vector is therefore a carrier of information about the direction of acoustic energy flow.

Parameters of the vector acoustic field were measured with the use of USP REGULAR type p-u probe which enables measurement of both the acoustic particle velocity in three directions and the sound pressure level, thus allowing direct determination of the sound intensity vector for selected points in space.

To determine the position of the probe tracking sphere, a stereoscopic infrared camera is used which tracks the probe's trajectory in three dimensions.

The actual position of the probe in space was calculated using VELO software with the Scan & Paint 3D module based on the results of the model fitting process carried out before each measuring session.

Two measurement surfaces were selected for detailed analysis of the acoustic vector field—a horizontal plane at half height of the tested physical model and a vertical plane at the distance of up to about 25 cm from the model. Due to physical limitations concerning identification of the intensity probe position by the infrared camera, only the upper half of the model was taken into account in the vertical plane.

10.4 AN EXAMPLE PHYSICAL MODEL OF A PHONONIC CRYSTALLINE STRUCTURE

With the use of the finite elements method (FEM), a model of a phononic crystal structure was developed with the lattice constant $a = 165$ mm comprising cylindrical scatterers with the radius of 70 mm. Under these assumptions, the fundamental Bragg's resonance frequency is (1039 ± 20) Hz. The band gap structure was calculated along three symmetry directions of the first irreducible Brillouin zone ΓX; XM, and MΓ based on a model developed with the use of the Bloch-Floquet theorem and numerically represented in terms of the FEM method.

Results of calculations indicate that the proposed arrangement of conventional phononic crystals are incapable of ensuring attenuation for all significant frequency components observed in examined noise sources. To increase the sound attenuation effectiveness, several variant solutions were considered, from among which a structure comprising six concentric annular resonators was chosen [Elford et al. 2011]. It was assumed that the proposed structure would be based on commonly available materials marketed in standardized dimensions (pipes, section bars, etc.). The physical model was made of PVC tubes with the density of $\rho = 1400$ kg/m³. A sketch of the physical model together with dimensions and the specifications of the utilized PVC components is shown in Figure 10.4.

The developed physical model comprised of three rows of resonators was placed in an acoustic chamber with properties close to those of the free acoustic field.

Figure 10.5 is a bar chart showing values of the sound insulation index *SI*, determined for the physical model in one-third octave bands of airborne sound [Radosz 2019]. In view of dimensions of the physical model, one-third octave bands centered in frequencies below 200 Hz were not taken into account.

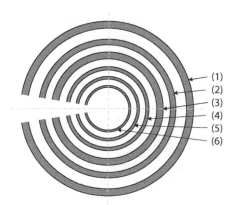

FIGURE 10.4 A dimensioned sketch of the physical model of a phononic crystalline structure. (1) Tube PVC-U PN 10 Ø140 × 5.4 mm, (2) Tube PVC-U PN 16 Ø110 × 6.6 mm, (3) Tube PVC-U PN 16 Ø90 × 6.7 mm, (4) Tube PVC-U PN 10 Ø63 × 3 mm, (5) Tube PVC-U PN 10 Ø50 × 2.4 mm, and (6) Tube PVC-U PN 16 Ø40 × 3.0 mm.

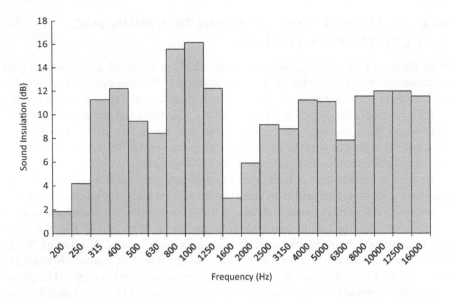

FIGURE 10.5 The sound insulation index of the physical model for airborne sounds. (From Radosz, J., *Appl. Acoust.*, 155, 492–499, 2019.)

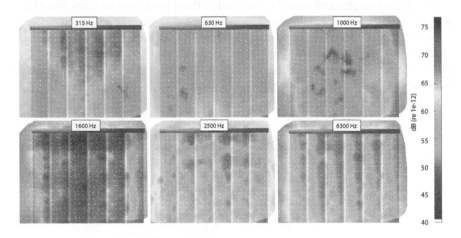

FIGURE 10.6 Distribution of the sound intensity vector in vertical measurement plane.

Figure 10.6 depicts the sound intensity vector distribution pattern in a vertical measurement plane. The results are presented in selected one-third octave bands with mid-band frequencies of 315 Hz, 630 Hz, 1000 Hz, 1600 Hz, 2500 Hz, and 6300 Hz.

The obtained sound intensity vector distribution patterns in the vertical measurement plane indicate a diversified effectiveness of sound insulation in the measurement plane, depending on position and frequency. For the band with the mid-band frequency of 315 Hz, one may observe the effect of diffraction and interference of sound waves which results in local sound intensity level variations of up to 20 dB.

The effect of interference of sound waves decreases with increasing frequencies. For the band with the mid-band frequency of 1000 Hz (encompassing the Bragg frequency), significant differences can be seen in the sound intensity level as high as 25 dB between the edges of the model and its central portion. For higher frequencies (6300 Hz), in view of lower ratio of the wavelength to the lattice constant, the effect of sound penetration through slits between resonators is clearly noticeable. For the band with the mid-band frequency of 1600 Hz, local sound intensity amplifications can be noted, especially in the central and upper portion of the physical model. For the outermost resonators on both sides of the model, significant sound damping was observed with the difference between the sound intensity level relative to the central and upper portion of the model of up to 20 dB.

Figure 10.7 shows the sound intensity distribution pattern in the horizontal measurement plane. The results are presented in selected one-third octave bands with mid-band frequencies 315 Hz, 630 Hz, 800 Hz, 1000 Hz, and 2500 Hz.

Just as in case of the vertical plane, the obtained sound intensity vector distribution patterns indicate diversified effectiveness of sound insulation in the measurement plane depending on position and frequency. For bands with mid-band frequencies 315 Hz and 630 Hz, it is also possible to observe the effect of diffraction and interference of sound waves which result in local variations of the sound intensity level of up to 20 dB. The effect of interference of sound waves decreases with increasing frequencies but is still noticeable up to the Bragg's resonance frequency. For the band with mid-band frequency of 1000 Hz, as in the vertical plane, one may observe significant differences in sound intensity levels reaching 20 dB between the edges of the model and its central portion. For higher frequencies (2500 Hz), the dominant effect of interference on the sound intensity distribution can be noted. For the band with the mid-band frequency of 1600 Hz, sound intensity amplification

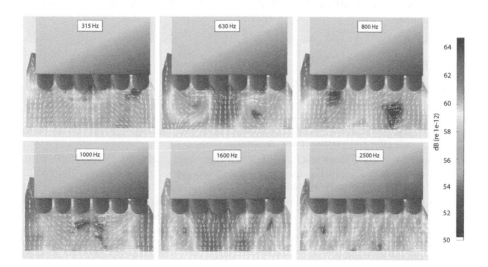

FIGURE 10.7   Distribution of the sound intensity vector in horizontal measurement plane.

was observed in the middle portion of the physical model. On the other hand, for the outermost resonators on both sides of the physical model, significant attenuation of sound was found with the sound intensity level difference relative to the middle portion of the physical model of up to 20 dB.

10.5 SUMMARY

The analysis of noise control effectiveness demonstrated by a physical model of a crystalline structure-based sound barrier with the use of the acoustic intensity probe presented in this chapter revealed diversified effectiveness of sound insulation in both the horizontal and vertical plane. The obtained sound intensity vector intensity patterns disclose local sound intensity level variations reaching, depending on the frequency, up to 25 dB, due to diffraction and interference of sound waves. For frequencies above 6300 Hz, the effect of penetration of sound through slits between resonators could be observed. Also the obtained sound pressure level distribution patterns indicate diversified effectiveness of acoustic isolation depending on frequency and position of the measurement point. Local changes in sound pressure level reduction can be as high as up to 30 dB.

From analysis of the literature of the subject [Lagarrigue et al. 2013; Morandi et al. 2016], it follows that there is a need to work out standardized methods for assessment of sound insulation performance of barriers based on phononic crystalline structures. In their above-quoted paper, Morandi and collaborators have proposed a method consistent with the standard EN 1793-6:2012 in which the barrier effectiveness assessment is carried out based on impulse response measurement in free field with the use of an appropriate time window to eliminate, among other things, spurious reflections. The sound intensity vector distribution patterns presented above suggest that the method based on measurements taken at nine points may turn out to be insufficient for the purpose of the sound insulation effectiveness assessment.

REFERENCES

Brillouin, L. 1946. *Wave Propagation in Periodic Structures*. 2nd ed. New York: Dover Publications Inc.
Caballero, D., J. Sanchez-Dehesa, and R. Constanza et al. 2000. Large two-dimensional sonic band gaps. *Phys Rev E Stat Phys Plasmas Fluids Relat Interdiscip Topics* 60(6 Pt A) R6316–9. DOI:10.1103/PhysRevE.60.R6316.
Castiñeira-Ibáñez, S., V. Romero-García, J. V. Sánchez-Pérez, and L. M. Garcia-Raffi. 2010. Overlapping of acoustic bandgaps using fractal geometries. *Europhys Lett* 92(2):24007. DOI:10.1209/0295-5075/92/24007.
Cervera, F., L. Sanchis, J. V. Sánchez-Pérez et al. 2001. Refractive Acoustic Devices for Airborne Sound. *Phys Rev Lett* 88(2):023902. DOI:10.1103/PhysRevLett.88.023902.
Chandra, H., P. A. Deymier, and J. O. Vasseur. 2004. Elastic wave propagation along waveguides in three-dimensional phononic crystals. *Phys Rev B* 70(5):054302. DOI:10.1103/PhysRevB.70.054302.

Chen, Y.-Y., and Z. Ye. 2001. Acoustic attenuation by two-dimensional arrays of rigid cylinders. *Phys Rev Lett* 87(18):184301. DOI:10.1103/PhysRevLett.87.184301.

Chong, Y. B. 2012. Sonic crystal noise barriers. PhD Thesis. The Open University. http://oro.open.ac.uk/44502/. (accessed January 28, 2020).

Elford, D. P., L. Chalmers, F. V. Kusmartsev, and G. M. Swallowe. 2011. Matryoshka locally resonant sonic crystal. *J Acoust Soc Am* 130(5):2746–2755. DOI:10.1121/1.3643818.

EN 1793-6:2012. Road traffic noise reducing devices—Test method for determining the acoustic performance—Part 6: Intrinsic characteristics—*In situ* values of airborne sound insulation under direct sound field conditions. 2012. European Committee for Standardization. Brussels, Belgium.

Fenech, B., and F. Jacobsen. 2005. Predicting the eigenmodes of a cavity containing an array of circular pipes. *J Acoust Soc Am* 117(1):63–67. DOI:10.1121/1.1836882.

Garcia, N., M. Nieto-Vesperinas, E. V. Ponizovskaya, and M. Torres. 2003. Theory for tailoring sonic devices: Diffraction dominates over refraction. *Phys Rev E Stat Nonlin Soft Matter Phys* 67(4 Pt 2):046606. DOI:10.1103/PhysRevE.67.046606.

Godinho, L., D. Soares, and P. G. Santos. 2016. Efficient analysis of sound propagation in sonic crystals using an ACA-MFS approach. *Eng Anal Bound Elem* 69:72–85. DOI:10.1016/j.enganabound.2016.05.001.

Goffaux, C., F. Maseri, J. O. Vasseur, B. Djafari-Rouhani, and P. Lambin. 2003. Measurements and calculations of the sound attenuation by a phononic band gap structure suitable for an insulating partition application. *Appl Phys Lett* 83(2):281–283. DOI:10.1063/1.1592016.

Goffaux, C., and J. P. Vigneron. 2001. Theoretical study of a tunable phononic band gap system. *Phys Rev B* 64(7):075118. DOI:10.1103/PhysRevB.64.075118.

Goffaux, C., and J. Sánchez-Dehesa. 2003. Two-dimensional phononic crystals studied using a variational method: Application to lattices of locally resonant materials. *Phys Rev B* 67(14):144301. DOI:10.1103/PhysRevB.67.144301.

Gonella, S., A. C. To, and W. K. Liu. 2009. Interplay between phononic bandgaps and piezoelectric microstructures for energy harvesting. *J Mech Phys Solids* 57(3):621–633. DOI:10.1016/j.jmps.2008.11.002.

Guild, Matthew D., V. M. García-Chocano, W. Kan, and J. Sánchez-Dehesa. 2015. Acoustic metamaterial absorbers based on multilayered sonic crystals. *J Appl Phys* 117(11):114902. DOI:10.1063/1.4915346.

Gupta, A. 2014. A review on sonic crystal, its applications and numerical analysis techniques. *Acoustic Phys* 60(2):223–234. DOI:10.1134/S1063771014020080.

Gupta, A., K.-M. Lim, and Ch. H. Chew. 2015. Design of radial sonic crystal for sound attenuation from divergent sound source. *Wave Motion* 55:1–9. DOI:10.1016/j.wavemoti.2015.01.002.

Hirsekorn, M., P. P. Delsanto, N. K. Batra, and P. Matic. 2004. Modelling and simulation of acoustic wave propagation in locally resonant sonic materials. *Ultrasonics* 42(1–9):231–235. DOI:10.1016/j.ultras.2004.01.014.

Hsiao, F.-L., A. Khelif, H. Moubchir, A. Choujaa, C.-C. Chen, and V. Laude. 2007. Waveguiding inside the complete band gap of a phononic crystal slab. *Phys Rev E Stat Nonlin Soft Matter Phys* 76(5 Pt 2):056601. DOI:10.1103/PhysRevE.76.056601.

Hu, X., C. T. Chan, and J. Zi. 2005. Two-dimensional sonic crystals with Helmholtz resonators. *Phys Rev E* 71(5):055601. DOI:10.1103/PhysRevE.71.055601.

Khelif, A., and A. Adibi, eds. 2016. *Phononic Crystals: Fundamentals and Applications*. 1st ed. New York: Springer.

Khelif, A., B. Aoubiza, S. Mohammadi, A. Adibi, and V. Laude. 2006. Complete band gaps in two-dimensional phononic crystal slabs. *Phys Rev E Stat Nonlin Soft Matter Phys* 74(4 Pt 2):046610. DOI:10.1103/PhysRevE.74.046610.

Khelif, A., P. A. Deymier, B. Djafari-Rouhani, J. O. Vasseur, and L. Dobrzynski. 2003. Two-dimensional phononic crystal with tunable narrow pass band: Application to a waveguide with selective frequency. *J Appl Phys* 94(3):1308–1311. DOI:10.1063/1.1557776.

Kittel, C. 2004. *Introduction to Solid State Physics*. 8th ed. Hoboken: Wiley.

Lagarrigue, C., J. P. Groby, and V. Tournat. 2013. Sustainable sonic crystal made of resonating bamboo rods. *J Acoust Soc Am* 133(1):247–254. DOI:10.1121/1.4769783.

Larabi, H., Y. Pennec, B. Djafari-Rouhani, and J. O. Vasseur. 2007. Multicoaxial cylindrical inclusions in locally resonant phononic crystals. *Phys Rev E Stat Nonlin Soft Matter Phys* 75(6 Pt 2):066601. DOI:10.1103/PhysRevE.75.066601.

Lardeau, A., J.-P. Groby, and V. Romero-Garcia. 2016. Broadband transmission loss using the overlap of resonances in 3D sonic crystals. *Crystals* 6(5):51. DOI:10.3390/cryst6050051.

Laude, V., Y. Achaoui, S. Benchabane, and A. Khelif. 2009. Evanescent Bloch waves and the complex band structure of phononic crystals. *Phys Rev B* 80:092301. DOI:10.1103/PhysRevB.80.092301.

Lee, H. M., K. M. Lim, and H. P. Lee. 2017a. Experimental and numerical studies on the design of a sonic crystal window. *J Vibroeng* 19(3):2224–2233. DOI:10.21595/jve.2017.17770.

Lee, H. M., L. B. Tan, K. M. Lim, and H. P. Lee. 2017b. Experimental study of the acoustical performance of a sonic crystal window in a reverberant sound field. *Build Acoust* 24(1):5–20. DOI:10.1177/1351010 × 16681015.

Liu, Z., X. Zhang, Y. Mao, Y. Y. Zhu, Z. Yang, C. T. Chan , and P. Sheng 2000. Locally resonant sonic materials. *Science* 289(5485):1734–1736.

Liu, Z. Y., C. T. Chan, and P. Sheng. 2002. Three-component elastic wave band-gap material. *Phys Rev B* 65(16):165116. DOI:10.1103/PhysRevB.65.165116.

Lu, J., C. Qiu, L. Ye et al. 2017. Observation of topological valley transport of sound in sonic crystals. *Nature Phys* 13(4):369–374. DOI:10.1038/NPHYS3999.

Martínez-Sala, R., J. Sancho, J. V. Sánchez, V. Gómez, J. Llinares, and F. Meseguer. 1995. Sound attenuation by sculpture. *Nature* 378(6554):241. DOI:10.1038/378241a0.

Miyashita, T. 2002. Full band gaps of sonic crystals made of acrylic cylinders in air – Numerical and experimental investigations. *Jap J Appl Phys* 41(5B):3170–3175. DOI:10.1143/JJAP.41.3170.

Montiel, F., H. Chung, M. Karimi, and N. Kessissoglou. 2017. An analytical and numerical investigation of acoustic attenuation by a finite sonic crystal. *Wave Motion* 70(SI):135–151. DOI:10.1016/j.wavemoti.2016.12.002.

Morandi, F., M. Miniaci, P. Guidorzi, A. Marzani, and I. Massimo Garai. 2015. Acoustic measurements on a sonic crystals barrier. In *6th International Building Physics Conference (IBPC 2015)*, ed. M. Perino, vol. 78, pp. 134–139. Amsterdam: Elsevier Science.

Morandi, F., M. Miniaci, A. Marzani, L. Barbaresi, and M. Garai. 2016. Standardised acoustic characterisation of sonic crystals noise barriers: Sound insulation and reflection properties. *Appl Acoust* 114:294–306. DOI:10.1016/j.apacoust.2016.07.028.

Oudich, M., Y. Li, B. M. Assouar, and Z. L. Hou. 2010. A sonic band gap based on the locally resonant phononic plates with stubs. *New J Phys* 12:e083049. DOI:10.1088/1367-2630/12/8/083049.

Peiro-Torres, M. P., J. Redondo, J. M. Bravo, and J. V. Sánchez-Pérez. 2016. Open noise barriers based on sonic crystals: Advances in noise control in transport infrastructures. In *Efficient, Safe and Intelligent Transport*, eds. J. V. Colomer, R. Insa, and T. Ruiz, vol. 18, pp. 392–398. Amsterdam: Elsevier Science.

Pennec, Y, J. O. Vasseur, B. Djafari-Rouhani, L. Dobrzynski, and P. A. Deymier. 2010. Two-dimensional phononic crystals: Examples and applications. *Surf Sci Rep* 65 (8):229–291. DOI:10.1016/j.surfrep.2010.08.002.

Phani, A., J. Srikantha, J. Woodhouse, and N. A. Fleck. 2006. Wave propagation in two-dimensional periodic lattices. *J Acoust Soc Am* 119 (4):1995–2005. DOI:10.1121/1.2179748.

Radosz, J. 2019. Acoustic performance of noise barrier based on sonic crystals with resonant elements. *Appl Acoust* 155:492–499.

Romero-García, V., J. V. Sánchez-Pérez, L. M. García-Raffi, J. M. Herrero, S. García-Nieto, and X. Blasco. 2009. Hole distribution in phononic crystals: Design and optimization. *J Acoust Soc Am* 125(6):3774–3783. DOI:10.1121/1.3126948.

Rubio, C., D. Caballero, J. Sánchez-Pérez et al. 1999. Existence of full gaps and deaf bands in two-dimensional sonic crystals. *J Lightwave Technol* 17(11):2202–2207. DOI:10.1109/50.803012.

Sainidou, R., N. Stefanou, and A. Modinos. 2002. Formation of absolute frequency gaps in three-dimensional solid phononic crystals. *Phys Rev B* 66(21):e212301. DOI:10.1103/PhysRevB.66.212301.

Sánchez-Pérez, J. V., D. Caballero, R. Mártinez-Sala et al. 1998. Sound attenuation by a two-dimensional array of rigid cylinders. *Phys Rev Lett* 80(24):5325–5328. DOI:10.1103/PhysRevLett.80.5325.

Sánchez-Pérez, J. V., C. Rubio, R. Martinez-Sala, R. Sanchez-Grandia, and I. Vicente Gomez. 2002. Acoustic barriers based on periodic arrays of scatterers. *Appl Phys Lett* 81(27):5240–5242. DOI:10.1063/1.1533112.

Sanchis, L., F. Cervera, J. Sánchez-Dehesa, J. V. Sánchez-Pérez, C. Rubio, and R. Martínez-Sala. 2001. Reflectance properties of two-dimensional sonic band-gap crystals. *J Acoust Soc Am* 109(6):2598–2605. DOI:10.1121/1.1369784.

Shakouri, A., F. Xu, and Z. Fan. 2017. Broadband acoustic energy confinement in hierarchical sonic crystals composed of rotated square inclusions. *Appl Phys Lett* 111(5):e054103. DOI:10.1063/1.4985230.

Sigalas, M. M., and E. N. Economou. 1992. Elastic and acoustic wave band structure. *J Sound Vib* 158:377–382. DOI:10.1016/0022-460X(92)90059-7.

Sigalas, M., and E. N. Economou. 1993. Band structure of elastic waves in two dimensional systems. *Solid State Commun* 86(3):141–143. DOI:10.1016/0038-1098(93)90888-T.

Tang, S. K. 2018. Reduction of sound transmission across plenum windows by incorporating an array of rigid cylinders. *J Sound Vibr* 415:25–40. DOI:10.1016/j.jsv.2017.11.027.

van der Aa, B., and J. Forssén. 2014. Scattering by an array of perforated cylinders with a porous core. *J Acoust Soc Am* 136(5):2370–2380. DOI:10.1121/1.4896566.

Vasseur, J. O., P. A. Deymier, A. Khelif et al. 2002. Phononic crystal with low filling fraction and absolute acoustic band gap in the audible frequency range: A theoretical and experimental study. *Phys Rev E Stat Nonlin Soft Matter Phys* 65(5 Pt 2):e056608. DOI:10.1103/PhysRevE.65.056608.

Index

Note: Page numbers in italic and bold refer to figures and tables, respectively.